# The Analysis of Organic Pollutants in Water and Waste Water

W. Leithe

Head of the Main Laboratory
The Austrian Nitrogen Company

*Translator*
STS, Incorporated
Ann Arbor, Michigan


*Consulting Technical Editor*
Nina McClelland, Director of Water Research
National Sanitation Foundation
Ann Arbor, Michigan

# FOREWORD

As a part of the worldwide effort in environmental protection, water conservation is receiving increasing attention in addition to that given to clean air. Large groups of the population, but particularly specialists in this field, are becoming increasingly concerned about the growing pollution of our water, often raising questions about adequately pure potable water supplies. This concern has resulted in the most diverse requirements and plans for measures to prevent this pollution in our water.

A necessary prerequisite for any consideration in this direction is to know the nature and concentration of pollutants existing in or introduced into the water; this knowledge must be provided by the analytical water chemist who thus performs a very important function in the field of water conservation. Since most residential and industrial water is discharged again in some form, but only after various impurities have been collected in it, it is particularly important to analyze waste water. This analysis is carried out by analysts in government agencies as well as in private industry.

Thus, increased efforts to prevent water pollution also lead to higher demands for its analysis. These demands result particularly from the diversity of new materials and active ingredients used in the trades and in industry, in agriculture and in the home. With increasing frequency, they reach the waste water from where they are introduced into the main sewer and the body of water receiving them. Many of these materials are harmful already in very low concentrations, necessitating particularly sensitive analytical methods.

The contaminants and pollutants of water are of inorganic as well as organic nature. While relatively few new analytical problems have arisen in the area of inorganic water analysis, organic pollutant analysis will continuously make more and more diverse demands on the analyst unless new methods, such as atomic absorption, become available. This applies not only to the overall determination and evaluation of organic pollutants, but also to individual substances with special effects in the sewer.

This material diversity of organic pollutants in waste water in particular also causes difficulties in developing standard methods of analysis, such as those which have proved useful in inorganic application. Moreover, the more diverse analytical conditions must be taken into account by a larger selection of analytical methods, and these must be known and judged by the analytical chemist. In this connection, a large number of newly developed methods of organic analysis are now available, especially gas and thin-layer chromatography, which have found acceptance in water analysis in the form of suitably modified procedures.

The author therefore considered it justified and desirable to investigate the field of analysis of organic water pollutants in a separate study and, with consideration of the characteristic nature of the subject, to make the results a companion to his book on "Analysis of Air and Its Pollutants" which was published by the same publisher.

The present book thus has the objective of compiling, screening and organizing pertinent information on analysis and methods as well as to offer special practical analytical experience for the benefit of analytical chemists, industrial chemists, doctors, public health specialists, and engineers working in the field of public water conservation. The above-noted practical experience was gathered by the author in his long years of activity as an analytical industrial chemist in Germany and Austria. The more important methods described in the domestic and foreign literature are presented in such detail that an analyst who is trained in the practical fundamentals of his specialty will be able to use them without reference to the original literature. As a rule several methods have been described for the more important pollutants to offer the analyst the possibility of making a choice based on the special conditions involved in his assignment. For less important procedures, only basic references could be supplied so that some insight into analytical characteristics and applications would at least be offered.

In addition to purely analytical methods and procedures, the marginal fields of water technology and hygiene are also lightly covered by way of an introduction. Therefore, the analyst who is not yet familiar with the subject is not placed in the position of being able to design an experiment appropriate for the needs and tasks involved, but being unable to understand and discuss the consequences of his results in the evaluation of a body of water or a given waste water. As a result, neither inadequate nor exaggerated conclusions will be drawn from the analytical results.

Accordingly, the book is organized into an introduction to the problems of organic water analysis with regard to their health and engineering aspects, together with pertinent literature, and a general characterization of analytical methods with special consideration given to chromatography separation methods, sampling, enrichment methods and odor testing. The specialized part deals first with the general determination of organic pollutants by elemental analysis and especially by oxidimetry and more recent instrumental methods. This is followed by the individual groups of pollutants with special polluting effects, their enrichment, isolation, separation and determination. These are particularly the cyanides, phenols, detergents, hydrocarbons, pesticides, urochromic agents, humic acids, pulp wastes, and polycylic aromatics. The appendix contains tabulations of purity requirements of waste water.

The author hopes that the present book will offer a useful tool to the water analyst and thus make a beneficial contribution to the urgent tasks of water conservation.

Wolfgang Leithe, Leonding near Linz

# CONTENTS

# 1. GENERAL DATA

## 1.1  Introduction

The statement that water, in addition to the air around us, is our most important element of life requires no special theoretical support for the human body consists of more than 60% water. The water eliminated in human metabolism must be continuously replaced through diet or fluid intake. Man can survive without food for several weeks, but in the absence of water death occurs in a few days.

Water serves not only for nutrition but is also in daily use for cleaning the body, as a laundering and rinsing agent for clothing, household articles, homes and streets. Still far larger quantities are employed in agriculture, industry and trade for the most diverse purposes, for energy generation, irrigation, cooling, rinsing and also as a raw material. It is therefore not surprising that the water supply is often of decisive importance for the life style of a population and the site selection of industry.

In addition to the *quantity* of needed water, its *quality* is also of greatest importance. Apart from a few properties which are directly characteristic of the $H_2O$ molecule, such as its temperature, this quality is essentially a consequence of its secondary components which influence its value in a positive as well as negative direction. Chemically pure water would be undesirable (at least in nature) since natural waterways would have no life without dissolved oxygen. Drinking water would not be healthy without a certain mineral content. However, the secondary components which *reduce* the consumption value of water are much more numerous.

These may be of *inorganic* nature, such as excessive water hardness or rock and soil suspensions which appear in the form of turbidity, particularly in spring water. Many pollutants originating from the home and industry are also of inorganic nature and can impair water quality by their quantity (alkali and alkaline earth salts) as well as by special toxic effects (for example, heavy metal ions). Recently, the pollutants of concern to the professional world and authorities in charge of water quality, because of their continuous increase in surface waters, are those derived from *organic* chemistry. This book is devoted to their analytical and quantitative determination.

## 1.2    Quantity and Origin of Organic Water Pollutants

### 1.2.1    *Water Requirements*

The distribution of water over the surface of the earth is extremely non-uniform. While many areas suffer from arid conditions and life must adapt to a water shortage there, and a frequent oversupply of water requires suitable protective measures in others, it is an advantage of our moderate climates that water is made available in sufficient and regular quantities.

The quantity of water available in a certain region results from the volume supplied as ground and surface water as well as the degree of precipitation. Average statistical values in the form of annual data in mm of ground cover are available with regard to the latter.

In West Germany, the annual average precipitation of 40 years (1891-1930) has amounted to 803 mm. Multiplied by a surface of 248,000 km², this results in a volume of about 200 billion m³/year. Approximately half of this evaporates again, while the balance remains in the country in the form of surface and ground water. A natural water supply of 25 billion m³ ground water and 75 billion m³ surface water is accepted as a basic value. At the present time, only a relatively small fraction of this quantity is consumed. In 1963, 3.7 billion m³ of water were used by various government units, while 9.6 billion m³ were used by industry. It is assumed that by the year 2000 the demand will double and surface water will therefore have to be used increasingly to cover the requirements.

In Austria (surface area of 84,000 km², average precipitation 1200 mm/year), the annual water supply amounts to 135 billion m³, consisting of 100 billion m³ from precipitation and 35 billion m³ by inflow from other countries.

The annual Austrian water requirements have been estimated at 1.5 billion m³ for industry, 0.5 billion m³ for domestic use and 0.1 billion m³ for agriculture. In 1969, 376 billion m³ were transported by the government.

## 1.2.2    Origin of Organic Pollutants

While rain and snow as a rule are free of organic pollutants, provided they have not absorbed such components from the fumes and waste gases of a polluted atmosphere, water absorbs various organic materials when it contacts and enters the soil. These are primarily compounds from humus- and peat-rich soil layers, some with a phenolic character with the characteristic of humic acids. If the soil is polluted by soluble organic wastes to such a degree that its adsorptive retention power is exceeded, the most diverse materials can be released to the water and can then be detected in the ground water. These may consist of fecal matter, components of human and animal cadavers, water-soluble components and decomposition products from improperly designed landfills as well as occasional petroleum residues resulting from transportation accidents or incorrect storage of fuels, or residues of pesticides used in agriculture and forestry. With percolation to greater depth and a sufficient residence time of the ground water in adsorptively active layers as well as with the accompanying activity of bacteria in the presence of sufficient oxygen, these pollutants may be degraded (mineralized) more or less completely.

After the ground water has become exposed in the form of surface water, it is endangered to an increased degree by organic pollutants. While natural pollutants, for example, leaves and entrained tree parts, as a rule play only a subordinate role, waste water of a domestic, agricultural and industrial nature in particular becomes a source of danger. As a result of the nearly complete sewer system the transported municipal and industrial water for the most part flows into surface waters but contains all pollutants which were introduced into it during its home, public and industrial use. Of the above-cited 9.6 billion m³ of industrial water transported in pipelines in West Germany, 1.3 billion remain unused, 5.9 billion are recycled as unpolluted cooling water, while 2.4 billion m³ were polluted.

If the organic pollutants remain within reasonable limits, they are degraded in their further transport in water by its bacterial flora with the use of dissolved oxygen (natural water treatment; see Chapter 1.2.5). In addition, increasing efforts must be made by industry to remove or destroy a part of the pollutants before discharge into the surface water, the "interceptor," by mechanical, biological or chemical methods.

### Distinction of the Origin of Organic Pollutants by Radiometric Age Determination

The radiometric age determination according to Libby makes it possible to distinguish whether the organic carbon in a water sample originated from synthetic products of petrochemistry and coal chemistry (natural gas, petroleum or coal raw materials) or from recent plant or animal matter. The concentration of radioactive $^{14}C$ found by radiometric analysis is the indicator. If this concentration corresponds to the average carbon content of the atmosphere, recent origin has been confirmed; if it is practically free of $^{14}C$ (half-life of 5568 years), a product from older geological formations is involved. Rosen and Rubin[261a], as well as more recently Kölle[168a], have made such analyses.

### 1.2.3  Residential Waste Water

An estimate of the volume of residential waste water and its organic pollutant content can be made with the use of mean values which, when multiplied by the population, result in the expected production.

According to Imhoff[144a], a domestic waste water volume without industrial components of 150 1/inhabitant/day serves as a basis for Germany. The following organic pollutants are assumed to be present in this:

| | mg/l | BOD$_5$ mg/l | g/inhab./day | BOD$_5$ g/inhab./day |
|---|---|---|---|---|
| Settleable suspended organic solids | 270 | 130 | 40 | 19 |
| Suspended organic solids | 130 | 80 ⎫ | 20 | 12 ⎫ |
| | | ⎬ 230 | | ⎬ 35 |
| Dissolved organic matter | 330 | 150 ⎭ | 50 | 23 ⎭ |
| | 730 | 360 | 110 | 54 |

The last value of 54 g BOD$_5$/inhabitant/day, occasionally referred to as the "Imhoff constant," has been verified repeatedly in recent years. Some authors are in favor of raising this value somewhat, while others see no need for a significant change.

Up until the last century, the disposal of residential and industrial waste, especially in cities, was extremely inadequate and frequently gave rise to catastrophic epidemics directly as well as indirectly by polluted drinking water. The greatest cholera epidemic in Hamburg in 1892 caused by polluted drinking water led to the death of 8,600 people. The construction of sewage systems in larger towns, particularly cities, which began soon afterwards, admittedly resulted in a considerable improvement of public health conditions at the site, but the pollutant load on rivers and lakes as the interceptors was increased significantly by this measure.

Efforts to treat residential and industrial waste water and construct the necessary plants began only approximately in the middle of the last century. The new knowledge and experiences in bacteriology acquired at that time served as a suitable foundation for a health evaluation of a body of water on the basis of the presence of pathogenic organisms and information was obtained concerning the nature and quantity of organic pollutants which make water a suitable nutrient for these organisms.

In any case, it became an acknowledged fact that the introduction of residential waste water into drinking water had to be prevented under all circumstances, but that pollution by pathogenic organisms of surface water serving for bathing and swimming, for example, also had to be avoided. Depending on the self-cleaning power of waterways, various treatment measures have proved to be necessary for residential waste water and industrial waste effluents of similar composition.

In 1968, 46 million of the approximately 60 million inhabitants of West Germany were connected to a sewer system. Twenty-two per cent of their waste water was discharged without treatment, 28% received only mechanical treatment and 50% was subjected to biological treatment. Since the end of the war up to 1968, $5 billion were spent by the government of West Germany for waste-water and sewage installations, including 1.4 billion for treatment plants. If 90% of the residential and industrial effluents are to be biologically treated, an additional $2.7 billion will be required.

### 1.2.4 Biological Water Evaluation According to the Saprobic System

(See also Liebmann[206], as well as DEV, 6th printing (1971), Group M 1-7)

The degree of pollution of a body of water can also be characterized by the number and nature of organisms living in it if their living conditions provided by the water quality are known. On this basis Kolkwitz and Marsson have developed their *saprobic system* in which they established the basic organisms for 4 qualities of water and classified them in lists of species. The system was later improved by Liebmann and it has become customary to record the results in colors on the streams shown in suitable geographical maps.

An attempt can be made to correlate the biological findings with the numerical results of chemical water analysis. The table

Table 1·2.4.—Saprobic zones according to Kolkwitz-Marsson

| | | Map color | No. of organisms/ ml | $BOD_5$ | COD (see Leithe) |
|---|---|---|---|---|---|
| Oligosaprobic zone (not polluted) | I | Blue | 100 | 3 | 1-2 |
| Intermediate zone | I-II | Blue/green | | | 4 |
| β-mesosaprobic zone (slightly polluted) | II | Green | 10,000 | 3-5.5 | 8-9 |
| Intermediate zone | II-III | Green/yellow | | | 11-18 |
| α-mesosaprobic zone (heavily polluted) | III | Yellow | 100,000 | 5.5-14 | 20-65 |
| Intermediate zone | III-IV | Yellow/red | | | 80-200 |
| Polysaprobic zone (grossly polluted) | IV | Red | >100,000 | 14 | |

lists the respective data on the basis of the $BOD_5$-value (according to Klotter and Hantge[166] and the COD-value (Leithe[199]).

Since a fairly long period of time is necessary for the development of the observed association in the water, a biological water analysis furnishes an average value for the water that has flowed during this time.

## 1.2.5   Biological Self-Cleaning

The possibility exists in ground water as well as surface water of degrading and detoxifying organic pollutants by a natural method by oxidative processes with the aid of bacteria and other micro-organisms—a process known as self-cleaning of water.

First the organic pollutants are partly dissimilated by bacteria (used for respiration), and in part they are transformed into new living bacterial substance (assimilated). A prerequisite for this process is the presence of bacterial species and strains having the capability for the necessary biochemical reactions. Such capabilities may be enhanced or newly formed in the course of time by natural selection processes. Thus, bacterial strains exist today which can metabolize and eliminate toxic substances in considerable concentrations, for example, phenol, formaldehyde, hydrocyanic acid, etc. The process is analogous to the development of strains with resistance to pesticides and drugs. Finally, the bacteria are also consumed by protozoa. This pathway finally results in the degradation of the organic pollutants into $CO_2$ and $H_2O$, while the nitrogen is converted into nitrate. The carbon dioxide formed and concentrated in the soil atmosphere can lead to an increased amount of dissolved calcium and magnesium carbonate and thus to an increase in water hardness.

This self-cleaning power is particularly important in ground water which is intended for drinking water. It is favored by air-permeable soil and slow flow velocities. However, suitable ground-water conservation areas must be developed in order to make a sufficiently large zone available to complete the elimination of organic pollutants and bacteria without producing a new burden if ground water is to furnish acceptable drinking water.

In surface waterways this cleaning power can also detoxify and eliminate considerable volumes of organic pollutants. However, the oxygen balance of a river or lake may not be excessively disturbed by this process. The oxygen demand for the biological oxidation of organic pollutants must be covered by a suitable absorption from the atmosphere. The oxygen concentration should not drop below 3-4 mg/l in order to maintain fish, protozoa and aerobic bacteria. U.S. standards are commonly 5 mg/l. If the oxygen is completely consumed by overloading of the water, the living organisms depending on oxygen die and the river enters a state of eutrophication. Toxins, particularly those of inorganic nature, also can interfere with biological self-cleaning.

## 1.3 Summary of Biological Treatment Methods for Residential and Industrial Waste Water

Residential effluents are subjected to *mechanical* treatment either immediately after discharge from a residence or at the inflow to the sewers in waste-water treatment installations using sand traps, rakes,,screens or settling tanks (Emsch septic tanks or Imhoff tanks). It is assumed that these installations trap about 30% of the total $BOD_5$ of the wastes.

The non-settleable or dissolved pollutants from residential effluents are removed further (about 40%) in *biological treatment plants* with the use of bacteria simulating or intensifying the biological degradation process taking place in nature in the rivers. Many industrial wastes of organic origin which are accessible to biological degradation can be treated in an analogous manner.

Numerous processes and installations have become known for this purpose and can be selected as a function of the population and construction conditions.

Two biological processes must be distinguished in particular:

1. Oxidative degradation of the organic substance into carbon dioxide, water and inorganic nitrogen compounds ($NH_3$, $N_2$ and possibly to the stage of nitrates);

2. Conversion of the organic matter into new bacterial material (sludge production) which is removed as a separate phase.

Process 1 requires much time and large-volume equipment and is, in fact, valid in theory only, while Process 2 offers a favorable possibility for the utilization and elimination of the formed sludge.

Two types of industrial devices predominate for biological purification: *trickling filters* and *activated sludge installations*.

### *1.3.1 Trickling Filters*

Trickling filters are large (5-50 m diameter) cylindrical tanks which are packed with beds of solids approximately the size of a fist (stones, clinkers). Recently, plastic gratings have been built into them in place of stones. The waste water is sprinkled on the beds with slowly revolving distributing nozzles and percolates through the aerated stone bed or grating. The stone surfaces or gratings are covered by a gelatinous film of living bacteria and other microorganisms (biological film) which absorbs the components of the waste water, partly consumes them for respiration and partly converts them into new bacterial biomass. With a low load, the formation of new biomass remains in narrow limits, so that its separation is often unnecessary. With a higher load, which may develop particularly on plastic gratings, the excess of sludge requires separation in a post-clarifying tank. Special mechanical aeration generally is not necessary if provision is made for adequate ventilation.

In previous decades trickling filters were the preferred biological purification process even for larger quantities of waste water. Their advantage consists of low maintenance and low energy consumption. Today they have been pushed somewhat into the background by the various activated sludge processes, but as a result of an increased efficiency obtained with new plastic gratings they appear to be becoming more popular again.

## 1.3.2  Activated Sludge Processes

The second important method for biological waste water treatment is the activated sludge process. The mechanically clarified waste water is conducted into open tanks into which large quantities of air are blown or intensively added by agitating systems (brushes or revolving paddles). The bacteria combine into freely suspended flocs. After a suitable detention time (average values of at least 4 h), the treated waste water is continuously drawn off and allowed to settle; a part of the activated sludge is returned into the aeration tank in order to intensify the action, while a part is removed as excess sludge.

The most diverse design variants are possible with this process as a function of the volume of waste water and its organic content, the desired effluent quality as well as the sludge production to be handled, in which the flow, dilution, sludge return, oxygen addition as well as the space arrangement of individual process zones can be widely varied and adapted to conditions. With a higher flow, a small installation may be sufficient if provisions have been made to dispose of the resulting larger quantity of sludge; conversely, the amount of sludge produced can be reduced or mineralized to such a degree that it becomes stabilized and odorless and can be disposed of more easily with the use of relatively large equipment, reduced flow and increased aeration.

Activated sludge processes are used particularly for municipal waste water treatment in which the desired purification effect can be adapted to the self-cleaning power of the interceptor. While interceptors with a high preload require intensive treatment (90-95% referred to $BOD_5$), waterways with a low load, large volumes of water, high flow velocities and therefore high capacities may not require more than a mechanical and partial biological treatment.

For smaller communities or plants with sufficient space, *oxidation ponds* are suitable, *i.e.*, annular ditches of about 1 m depth in which the waste water is agitated and aerated with the use of rotating rolls. They can be operated in batches or continuously.

An intermediate process between trickling filters and activated sludge installations is that using immersion trickling filters in which a large number of closely spaced concentric disks of 2-3 m diameter rotate slowly and become covered with an activated biological film. These filters are called rotating disk filters in the U.S. and are used here only experimentally.

Apart from the biological treatment processes described above, smaller volumes of waste water or those with a low load can also be disposed of and made harmless with simpler systems. Such processes, for example, consist of land treatment on septic fields, natural filtration with simultaneous irrigation as well as wastewater ponds which may also serve for fish breeding under certain conditions. However, because of their relatively small capacity, these processes are only of local significance. Individual aerobic systems are also permitted by some regulatory agencies in the U.S.

### 1.3.3    The "Tertiary Treatment Stage"

Recently, particularly in the case of waste water production in the immediate vicinity of lakes with a significant tourist business, a so-called tertiary treatment stage is considered necessary beyond the biological stage. It consists of the removal of nitrogen and phosphorus compounds from the effluent which can produce increased algae growth and thus damage of a lake due to excessive fertilization. The phosphates can be precipitated with iron or aluminum salts, possibly as a part of biological treatment. The nitrates can be eliminated by an anaerobic intermediate stage with bacterial consumption of the nitrate oxygen in the form of $N_2$ or $N_2O$ (nitrous oxide). Where indicated, however, it will be preferred to divert the waste water from the lake by a peripheral sewer system.

### 1.3.4    Sludge Disposal

An important partial problem of waste water treatment consists of the disposal of sludge, particularly when most of the organic pollutants have been converted into sludge according to process step 2 (see above). This material therefore consists of fresh sludge from the preliminary settling step and the surplus sludge from the activation process.

In its raw form sludge cannot be disposed of because of its tendency to become septic and cause an odor nuisance. It can be stabilized under aerobic conditions by further treatment with air, which largely mineralizes most of the organic matter as a result of respiration. If the volumes are not excessively large, it can be spread on drying beds in this form.

With larger volumes of sludge, anaerobic digestion in suitable digestion tanks at elevated temperature for several days is a more frequent process. The resulting combustible gases

$(CH_4 + CO_2)$ are utilized for heating and energy recovery in the treatment plant. With complete digestion, the sludge solids are reduced to about $1/3$ of the original quantity and the water content to about 90%. The foul odor associated with $CH_4$ is eliminated.

The most favorable step would be to pass digested sludge on to agricultural uses, if it became possible to solve the transportation problem, for example, the availability of suitable large-volume vehicles. However, the demands of public health specialists must be kept in mind who require destruction of pathogenic organisms by pasteurizing at 70°. It is also possible to compost sludge together with residential refuse.

Sludge can also be dewatered further by mechanical methods with filters or centrifuges. Dewatering can take place with the application of heat, for example, in drying drums in a cyclic process in which the wet raw sludge is charged onto the surface of already dewatered material. Subsequently, it can be utilized as a soil builder or can be incinerated. It is true that inexpensive heat sources must be available for an economical evaporation of such large quantities of water.

### 1.3.5 Biological Treatment of Industrial Effluents

Many industrial effluents are similar to residential waste water in qualitative composition, such as those from dairies, sugar refineries and canning plants. With suitable metering, their treatment in biological plants presents no basic difficulties, particularly since the respective pollutants require no special measures regarding bacterial flora and only a certain degree of caution is necessary with regard to loading. Load peaks can be compensated for by larger holding tanks in a plant.

Surprisingly, many organic substances which do not occur in nature or are only hypothetical intermediates in biological cycles have proved to be readily biodegradable if provisions for an effective bacterial adaptation have been made in the treatment plants by a sufficient adaptation time, possibly by inoculating with sludge from plants already operating on the same principle. Compounds which are known to be disinfectants and bacterial toxins, for example, phenol and formaldehyde, can also be degraded in suitable dilution. In such cases, however, favorable biological conditions are necessary, for example, with regard to the presence of the most important inorganic plant nutrients, *i.e.,* nitrogen and phosphorus, prevention of toxic concentration ranges, as well as the absence of inorganic toxins. In this respect, common treatment of industrial and residential waste water is particularly attractive, since the latter furnishes not only an enrichment of bacterial flora and a supply of nitrogen and phosphorus but also suitable dilution.

In many chemical plants, biological treatment plants of gigantic dimensions (several million population equivalents) are already in operation or in the planning stage in order to provide relief for the already highly loaded rivers. Costs of such plants add up to the billions of dollars.

The composition of the waste effluents from the chemical industry proper is as diverse as its products. Water-soluble wastes from large-scale production, particularly for the manufacture of plastics raw materials, pesticides, detergents and the like, are particularly important.

Quantitative data concerning the volumes of pollutants to be expected in industrial production are difficult to obtain, since the quantities referring to a production unit differ highly from one plant to another depending on the manufacturing process, raw materials, process control and waste disposal measures. The government is exerting increasing pressure on industrial plants to reduce their waste production as far as possible by modifying their processes, recycling, etc.

Special treatment of industrial waste effluents by methods of chemical process technology is becoming increasingly urgent in order to preserve the interceptors.

## 1.4   Chemical and Physical Treatment Methods of Industrial Waste Effluents

The industrial measures for water conservation consist of steps to prevent or at least reduce the production of pollutants or their introduction into the waste effluents, on one hand, and of measures to remove pollution from the waste water and to discharge it into the interceptor in suitably treated form on the other.

Various internal plant measures for the prevention of water pollution have been recommended frequently:

1. Preventing the formation of byproducts which would otherwise reach the waste effluents by modifying the process. For example, a more suitable catalyst may make a reaction more specific.

2. Choice of a more suitable raw material which does not contain or does not allow the formation of a pollutant which would otherwise be discharged with the waste effluent.

3. Cycling of wash water. The enrichment of the extracted material may permit its isolation and frequently its further utilization. The residual solutions can be recycled to the process.

### 1.4.1  Organic Components of Industrial Waste Effluents

Organic Components of Industrial Waste Effluents

| Plant | Waste Effluent | Composition |
|---|---|---|
| Mines, ore treatment plants | Mine and wash water | Humus, coal sludge, flotation agents |
| Foundries | Blast-furnace gas wash water | Cyanides, phenol, tar components, coal sludge |
| Iron and steel processing | Aging, cooling and pickling solutions, rinse water | Wetting agents and lubricants, cyanides, inhibitors, hydrocarbons, solvent residues |
| Coal production, coking plants | Wash water, ammonia liquor | Humus, coal particles, cyanides, rodanines, phenols, hydrocarbons, pyridine bases |
| Wood charcoal production | Gas scrubbing water | Fatty acids, alcohols, particularly methanol, phenols |
| Petroleum industry | Drilling water, gas scrubbing water, rinsing water, acid condensates | Oil emulsions, naphthenic acids, phenols, sulfonates |
| Sulfite pulp | Sulfite liquor | Methanol, cymol, furfurol, soluble carbohydrates, lignosulfonic acids |
| Soda (sulfate) pulp | Vapor and digester condensates | Mercaptans and sulfides, alcohols, terpenes, lignin, resinic acids, soluble carbohydrates |
| Rayon and cellulose | | Xanthogenates, alkali hemicelluloses |
| Paper manufacture | | Resinic acids, polysaccharides, mucins, cellulose fibers, flotation agents |
| Textile industry | | Scouring and wetting agents, leveling agents, sizers, desizing agents, fatty acids, finishes, Trilon (nitrilotriacetic acid), dyes |

### Organic Components of Industrial Waste Effluents—Continued

| Plant | Waste Effluent | Composition |
|---|---|---|
| Laundries | | Detergents: carboxymethyl-cellulose, enzymes, optical brighteners, colorants; soil: protein, blood, cocoa, coffee, carbohydrates, emulsified fats, soot |
| Leather and tanning industry | | Protein degradation products, soaps, tanning agents, emulsified lime soap, hair |
| Natural glue and gelatin | Alkaline wash water, vapor condensates | Protein degradation products emulsified fats and lime soaps |
| Sugar refineries | Beet wash water, chip press water, vapor condensates | Sugar, plant acids, betaine, pectin and other soluble plant components |
| Starch plants | Rinsing and wash water | Water-soluble plant components (protein compounds, pectins, soluble carbohydrates) |
| Dairies | Rinsing water, butter churning water | Milk components (protein, lactose, lactic acid, fat emulsions), washing and rinsing agents |
| Grease and soap factories | Washing and kneading water | Glycerine, fatty acids, fat emulsions |
| Canning factories | Washing and blanching water | All types of soluble plant components |
| Sauerkraut factories | | Lactic, acetic and butyric acids, carbohydrates, other soluble plant components |
| Beer breweries | Malt, soaking and steeping water, yeast wash water, rinse water | Water-soluble plant components, beer residues, rinsing agents |
| Fermentation industry | Slops | Fatty and amino acids, alcohols, unfermented carbohydrates |
| Slaughter houses | | Blood, water-soluble and emulsified meat components, fecal matter |

## 1.4.2  Industrial Processes

By far the predominant part of the industrial water used by large-scale chemical plants is not polluted in the process but serves only for cooling purposes, for example. At favorable ambient temperatures, it can be cooled in large cooling towers and reused. In older plants, these unpolluted effluents are discharged together with surface water, fecal wastes and polluted industrial water into a common mixing sewer. In order to be able to collect the polluted effluents in not too dilute form and subject them to a treatment process, a separate discharge of polluted effluents from the large quantities of unpolluted or only slightly polluted waste water is necessary, *i.e.,* construction of a separate sewer system which incurs great difficulties in construction and costs in existing older facilities.

For the separation of mechanical contaminants of inorganic and organic nature, the most diverse devices are in use: depending on size and cross section of the particles, rakes, screens or filters, or as a function of the specific gravity, sand traps, settling tanks or centrifuges. Liquids with a density of less than 1 which are not dissolved in water are separated in mechanical oil and gasoline separators. Standards are available for their construction (DIN 1999 for gasoline separators and DIN 4040 for grease separators). The detention time of the waste effluent in these separators should be at least 3-30 min.

Toxic contaminants can be recovered by *extraction.* The most important is the extraction of phenols from coking and low-temperature carbonization effluents. It is performed most frequently with captive benzene and occasionally with special extractants (phenosolvan, butylacetate).

The separation of fine pollutants which do not settle readily can be facilitated by *flocculating aids.* Inorganic flocculating aids consist of aluminum and particularly of iron salts, and among the latter, preference is given to Fe(III) salts; these can be obtained from iron(II) sulfate, which is available in large quantities for the surface treatment of sheet metal and wires with sulfuric acid and is then converted with chlorine and recently also with atmospheric oxygen and activated carbon.

Organic water-soluble high polymers (for example, polyacrylates, polyacrylamides, polyethylene oxides, carboxymethylcellulose), which are marketed under tradenames such as Sedipur, Praestol, Separan, Magnifloc, etc., are becoming of increasing importance. They must be correctly metered so that a residual excess in the water does not become a pollutant itself. These materials with colloidal activity form agglomerates with the suspended pollutants which carry the opposite charge and settle readily and are therefore easily separated. Modern stirred tanks are frequently used for this purpose in which flocculation and separation can be combined (Gyromat, Accelator).

The effect of flocculation and separation of solid particles may be increased under certain conditions by flotation effects which are produced by the injection of air after addition of a precipitant, usually a surfactant. The material to be removed is enveloped in the form of foam and forced to rise. An example of this is the separation of fine cellulose fibers in the paper industry by the injection of air after addition of a surfactant. Again the danger exists here that an excess of surfactant will reach the discharged effluent. The effect of flocculating aids and flotation agents is not limited to the separation of insoluble solids but can also serve for the removal of colloidal substances or true solutes (for example, detergents).

For a chemical precipitation of dissolved pollutants, it is also possible to use additives which undergo a chemical reaction with the dissolved materials with the formation of water-insoluble precipitates. Neutralization is frequently also desired if the isoelectric point of the pollutant where solubility is at a minimum is near a pH of 7. As an example we may cite the introduction of carbon dioxide in the form of flue gases into highly alkaline protein-liming effluents from canning plants. Conversely, some acids can be precipitated as calcium salts by the addition of calcium. At a suitable pH value, cyanides can be treated with iron salts and precipitated and separated in the form of Berlin blue [iron (III)-cyanoferrate(II)].

Practical examples of flocculation and precipitation methods are described by Dietrich[62a] as well as Oehler[242a].

*Waste water treatment by adsorption.* Many solvents adsorb certain materials on their surface from solutions. These materials can then be removed again from the adsorbent by suitable measures (for example, extraction) and under certain conditions can be reused. Activated carbon, which retains its adsorptive power even in the presence of water, is particularly suitable for aqueous solutions. Examples of this process are the removal of colorants from spent dye solutions as well as the removal of trace pollutants in drinking water, for example, undesirable osmophores, insecticides, mineral oils, etc., with the aid of activated carbon filters.

The specified properties and testing of commercial activated carbons are listed in DIN Specification 19,603 (1969). The efficiency is tested by adsorption tests with phenol solutions of known concentration.

Free chlorine can also be removed from water with activated carbon.

*Ion exchangers.* Ion exchangers, which are more commonly used to remove metal ions, can occasionally also be employed for the separation of organic acid pollutants, for example, cyanides, phenols and organic acids. Suitable anion exchangers for the purpose are synthetic resins with built-in basic groups (amino and quarternary ammonium groups). They are converted into the OH-form by means of strong bases and thus bind the mentioned acid residues from waste water. Strong alkalis elute the bound anion in concentrated form and the exchanger is regenerated for reuse.

*Steaming of waste water.* Many industrial waste effluents contain volatile organic compounds in dissolved or emulsified form. These can be driven out by steaming at elevated temperature, for example, with submerged flame burners, with or without injection of a gas, recovered by condensation or pyrolyzed on catalysts. This is particularly favorable for those hydrocarbons which are not biodegradable. A part of the heat supplied to the waste water during heating can be transferred to new batches by suitable heat exchangers. The prevention of air pollution and odor nuisances must be kept in mind in all such processes. The demands for clean air are no less severe than those of water conservation.

*Oxidation and combustion processes.* Easily oxidizable pollutants of low concentration can be eliminated by suitable oxidants, such as chlorine, chlorine dioxide or ozone. An important application for this process is the oxidation of cyanides in highly alkaline solution (pH $> 11$) by treatment with chlorine which converts them into less toxic cyanates. In the treatment of drinking water with chlorine or ozone for disinfection (destruction of pathogenic organisms) oxidants are also used for the oxidation of organic pollutants. Occasionally, for example, in the presence of phenol, this results in the formation of chlorinated phenols with an extremely intense odor and taste. Their formation remains absent if chlorine dioxide or ozone is used instead of chlorine.

Naturally, nonvolatile pollutants can be obtained as a residue by evaporation of the water—a process which can be used only for concentrated effluents because of the considerable evaporation costs. If no other use appears feasible, the residues can be incinerated.

Such concentrated waste effluents consist particularly of the sulfite waste liquors in pulp production. About one ton of solids (lignosulfonic acid, pentosane, sugar) is produced per ton of pulp in the form of an approximately 10% solution. The effluents from the calcium bisulfate process, in particular, present great difficulties during evaporation because of scale buildup which is difficult to control. It is of greater advantage to use magnesium bisulfate solutions which do not cause scale buildup during evaporation: during heat treatment, the magnesium remains in the form of oxide suspended in water and thus allows an isolation of sulfur dioxide from the combustion gases with an extensive recovery of chemicals used for wood digestion (see also Hornke[139a]).

Another possibility for the removal of organic pollutants in waste water by oxidation is offered by wet oxidation with air at elevated pressure (50-150 atm) and elevated temperature (200-300°) for the recovery of heat in the form of steam. This process (Zimmermann process) was considered very promising for the treatment of pulp waste effluents but these hopes were not fulfilled. Better results were obtained in the incineration of wet sludge.

Occasionally, particularly with high concentrations of easily combustible components such as those formed in oil refineries, for example, the waste water can be atomized with compressed air and burned in large incinerators maintained at about 1000° (possibly together with used oil or combustible refinery waste gases). A part of the heat generated can be recovered in the form of steam.

# 2. GENERAL METHODS

## 2.1 Objectives and Purposes of Organic Water Analysis

Evaluation of the quality of a water sample is based on a knowledge of the nature and quantity of its contaminants, of which the organic pollutants are the subject of this book.

First, presence of many such impurities can be detected by a simple sensory test. The materials distributed in the water in the solid or as a second liquid phase are particularly evident. These are suspensions of solids with the nature of soil, occasional oily emulsions or foams formed during shaking, or thin, often iridescent films adhering to the surface. Dissolved impurities may be identifiable by a yellowish color, marked taste or odor, the latter especially on heating or boiling.

Their odor reveals even the slightest traces of impurities which as such are not yet toxic, but nevertheless remain indices of the presence of residential or industrial waste effluents (see chapter 4.1).

With the progress of hygiene and technology, it soon became apparent that a sensory test by no means allows the determination of all important organic impurities. The chemistry of water analysis therefore accepted this task. Numerous chemical and physical analytical methods came to be used; methods of determination were developed for increasingly diverse organic substances with concentrations which became lower and lower but nevertheless remained important in terms of hygiene or technology.

In this context the practice of water conservation continuously presented new problems for chemical analysis in order to allow a rapid, reliable and specific identification and determination of impurities, even in extremely low concentrations. New developments in the field of chemical analysis, *e.g.*, atomic absorption in the inorganic sector or the various chromatographic techniques for organic pollutants, found very fruitful applications in water analysis and beyond this led to effective measures for water conservation. Instrumental analysis also received notable impetus by the need for continuous automatic water analyses. Developments are still proceeding in full swing in all of these fields and the progress of analytical methodology has by no means stopped.

Water analysis is performed with water samples of very different degrees of purity. Spring and ground water, especially when used for drinking purposes as well as in the food industry, for example, must have a low organic impurity concentration and its determination is particularly important in these fields. Surface waters usually contain a higher load of impurities. In residential and industrial waste effluents the content of organic pollutants may be greater by several orders of magnitude.

The analyst, then, has many important tasks. In the case of drinking water, the water quality determination already begins during the design and construction of the recovery installation, and later it serves for continuous control during use as well as to verify the efficiency of treatment processes (aeration, clarification, chlorination, ozonation), as well as for the occasional determination of unexpected contaminants (entry of surface water or of residential or industrial pollutants, mineral oils, etc.). Drinking water is subject to legislated controls (the Food Act, West Germany) and the purity specifications established in this legislation (see also DIN 2000 and 2001).

Bacteriological and biological assays are closely related with chemical water analysis. They require special training and therefore are usually carried out by specialists with specific experience in this discipline. They are not discussed in this book.

The relationship of the bacteriological findings with an organic-chemical water analysis results from the fact that organic pollutants furnish the nutrients for bacteria. Thus, bacteria together with the oxygen dissolved in the water represent the agent which determines the fate of organic water pollutants under natural conditions. Organic water pollutants are particularly important when they serve as the nutrient substrate for pathogenic organisms of all kinds.

Various sectors of industry, especially the food and pharmaceutical industry, set the same or even more severe bacteriological and chemical requirements for their industrial water as those of drinking water, and this includes color and any aromatic or flavor-imparting contaminants. As a rule, swimming pools also must use drinking water quality. In these, analyses for special contaminants (urea, urochrome) become necessary.

For evaluation of drinking water from the hygiene standpoint

often must be determined with high accuracy. Concentration fluctuations can be determined by taking water samples at appropriate frequency or, when available, by continuous instrumental analyses.

## 2.2.2  Required Accuracy

The term "accuracy" of an analytical method refers to its "sensitivity," its "reproducibility" and its "correctness."

The sensitivity of an analytical method corresponds to the minimum concentration (limit of detection according to Feigl) or the lowest concentration (concentration limit) of a substance to be determined in water, which thus is detectable with a specified reliability. It is obtained with consideration of the blank value of the method when the analysis is performed with all reagents, but with pure water instead of the water sample under otherwise identical conditions. The blank value may be the end point of a titration, or may result from weighing or spectrophotometric determination or the like. On the basis of a fairly large number of such blank value determinations, their standard deviation is calculated according to the following formula:

$$s = \pm \sqrt{\frac{(X_i - \bar{X})^2}{N - 1}} \quad \begin{array}{l} X_i \\ \bar{X} \\ N \end{array} \quad \begin{array}{l} X_i = \text{single values} \\ \bar{X} = \text{mean of all single values} \\ N = \text{number of single values} \end{array}$$

The mean blank value represents the base or reference value. If repeated analyses of a water sample show a value $X \pm s$, the presence of a given substance has been confirmed with 68% probability. If it is higher than the mean blank value by 2 s, the probability increases to 95% and with 3 s to 99.7%.

The "correctness" of an analytical method is confirmed either by parallel determinations by a "standard method" which is recognized as being especially reliable or, more safely, on the basis of test solutions of weighed quantities of the given substance in water which, if necessary, may also contain certain impurities in a concentration as anticipated in the practical sample.

The necessary sensitivity of a method to be applied depends on the problem to be solved. If a question of pollution is involved, the limit concentration of the method of detection should be approximately one power of ten below the known limit value of the damage produced by the given substance. In this type of problem, a standard deviation of $\pm 5\%$ of the test value represents a suffi-

cient reproducibility; this can be easily realized by conventional titrations, spectrophotometry or polarography. If very inaccurate data are known about the cited limit value of the pollutant, it is useless to attempt to attain a high percentual reproducibility. Reproducibilities of better than $\pm 1\%$ of the total value will therefore be rarely required in water analysis and determinations with a relative standard deviation as high as $\pm 10\text{-}20\%$ will be sufficient. Basically, it is preferable to save time and costs rather than to make efforts toward obtaining a higher than necessary accuracy.

### 2.2.3    Specificity

The *specificity* of a method is very closely related to the question of the correctness of an analysis. This factor is particularly important in the choice of a method and must be in accord with the purpose and more specific circumstances of the analysis.

Situations exist in which the detection of a very specific substance is required, for example, when damage produced by a pesticide demands the determination of its origin and of the polluter. In other instances, the determination of a group of substances of identical effect will suffice as in the case of detergents, if a distinction between biodegradable (soft) and bioresistant (hard) agents is not necessary.

In many cases, the most extensive determination of the sum of all organic substances in a water sample is even desirable, while in others the determination will have to be limited to the substances which are biodegradable by bacteria in a set period of time.

As a rule, higher specificity demands result in increased investments in analytical equipment. For example, if each member in the group of phenol-containing substances is to be determined separately, group reagents will not suffice and separation techniques, especially from the chromatography sector, will have to be available. The specificity requirement will be particularly difficult to satisfy when individual substances as carriers of particularly toxic activity without special chemical reactivity from a group of chemically and analytically similar compounds are to be investigated separately, as for example, in carcinogenic cycloaromatic hydrocarbons.

### 2.2.4    Time Requirement

The time requirement is of dual significance in the present context: first in regard to the necessary personnel which, however,

becomes apparent only when actual work is done with the sample during this period of time (manipulation time), while simple down-time is of less importance provided the working schedule is suitably organized.

Of much greater importance is the period of time from the start of analysis up to availability of the result in view of the consequences involved. If instantaneous introductions of pollutants may have acutely dangerous consequences, which must be quickly counteracted, only rapid methods with the use of frequent and preferably continuous sampling are applicable.

An important factor in the choice of method concerns the question whether a request involves single or only a few analyses or whether a series of these is to be performed, possibly for a longer period of time. As a rule, provisions for mechanization and automation as well as instrumental methods requiring time-consuming calibration procedures will save time only when large series of tests are required.

### 2.2.5    Costs of Analyses

In principle it is always appropriate for the analyst to inform the requester of the anticipated costs of an analysis in order to avoid that the latter will not be commensurate with the value of the desired information. As a rule, generosity in making funds for water analysis available will be guaranteed today by the widely accepted recognition of the pertinence and importance of water conservation. Consequently, the analyst will not need to shy away from higher procurement costs for apparatus which furnishes adequate information of a sufficient number of samples or allows a saving in manpower, which is usually necessary today. At the same time it must be considered that the costs of analyses always represent only a small part of the investments in water conservation, while inadequate analytical bases may lead to the neglect of necessary protective measures or to incorrect investments, which may amount to a multiple of the possible savings in analysis costs.

# 3. PRELIMINARY STEPS

## 3.1 Water Sampling

Many literature data as well as a broad range of apparatus offered by equipment manufacturers exist for water sampling. If organic components are to be analyzed, especially in industrial effluents, a few aspects are of greater importance than others:

1. Losses of volatile constituents during agitation or in the presence of larger amounts of air.

2. Possibility of bacterial degradation or chemical modification, *e.g.*, oxidation or hydrolysis, of organic components in the period between sampling and analysis.

3. Presence of organic impurities as a second phase in nonuniform distribution in the water.

In the choice of vessels for water samples to be tested for organic pollutants, great caution is indicated with the use of plastics. There is danger that these may release organic constituents into the water sample as well as adsorb these on the plastic vessel wall from the water, as has been reported by Quentin and Huschenbeth[254] in connection with pesticide losses. According to Hellmann[131] this also applies to hydrocarbons. Consequently, glass vessels with glass stoppers will be preferable in spite of their disadvantages. Naturally, all sample vessels must be carefully cleaned before use; no residues of organic cleaning agents may remain in them.

The importance of correct and representative sampling is known to the analyst but not always to all persons connected with this task. Therefore, if the water sample cannot be taken by the analyst himself or by his trained representative, sampling should take place at least after consulting the analyst in order to obtain his instructions with a full indication of the purpose of an analysis and all conditions involving the sample.

If sampling presents difficulties in stagnant material and water, these become even greater when samples are to be taken from moving materials of variable composition, as from flowing water or continuously produced production effluents under varying production conditions. The person taking the sample must therefore be just as informed as the analyst with regard to the purpose of the analysis on the one hand and all local and time conditions on the other hand, and must record these on the sample vessel. Suitable provisions must be made for the possibility of toxins in industrial effluents ($H_2S$, cyanides).

In this regard it is first of all necessary to establish the situation which the water sample is to identify whether instantaneous conditions with all instantaneous anomalous influences are to be reflected or whether a characterization of a normal state, *e.g.*, in a sewer, is involved.

In the case of moving water, the season, the instantaneous and previous weather conditions, course of the water, location of waste discharge points and the waste currents formed by them, as well as the entry of tributaries of different composition are of interest. These factors also determine the distance from shore as well as the water depth from which the sample is to be taken.

In the investigation of residential waste water, the expected fluctuations in composition as a function of time of day, working time of those in the home, weekends, presence of visitors or guests, transients, etc. must be taken into account. Similar considerations apply to sampling from biological treatment installations.

If the condition of a given body of water or of a waste water and thus its organic impurity content can vary in brief time intervals, samples must be taken at suitable frequency or continuously. A decision must then be made whether these samples are to be tested separately or whether a pooled sample covering a longer period of time will be sufficient. The quantitative ratio of a pooled sample should be adapted to the water flow in the respective time intervals, *i.e.*, it should be in quantitative proportion. In this case, a procedure must be chosen which is controlled by the quantity of outflow per unit of time.

The location of sampling sites for industrial effluents depends on whether a certain location is to be investigated, *e.g.*, the waste effluent from a special production process or whether a representative uniform mixture for the entire operation is to be sampled and analyzed.

As mentioned earlier, the demands for obtaining a representative average are particularly difficult to satisfy if the contaminant is discharged in the form of a second phase, be it solid or a second liquid phase. Then it will often be most advisable to sample and analyze each phase separately and to subject the quantitative ratio of the two phases to separate study.

The permissible period of time between sampling and analysis as well as the necessary preservatives differ from one case to another and are to be stated for the individual substances. Storage must always be in the absence of light in order to prevent algae growth.

A new automatic sampler PNG 12 (Braun in Melsungen) to draw mixed samples in proportion with time and volume is now offered. It meters variable volumes of liquid (20-500 ml) at optional time intervals (1-30 min) into 12 one-liter receivers of plastic or glass. The collection period per vessel can be set at time intervals ranging between 1 h and 1 day. The unit can be installed in the field.

## 3.2  Enrichment of Organic Water Pollutants for Analysis

Although many analytical methods are so sensitive that they can be performed directly with the water sample, enrichment of the substances to be determined in a larger volume of sample is sometimes necessary. This applies especially to extremely low pollutant concentrations below the limit of detection of a given reaction. A number of methods are available for this purpose and these will first be described on the basis of general aspects.

*Adsorption methods.* Activated carbon is preferred as an adsorbent for aqueous solutions. The procedure with subsequent extraction from chloroform (CCE method) is described in section 12.2. Several hundred liters of sample water can be used.

The special advantage of activated carbon adsorption consists of the high degree of enrichment since relatively large quantities of substance can be isolated from a large volume of water even with very low pollutant concentrations. However, it is a disadvantage that complete elution of the adsorbates is not always possible and that the latter are often chemically modified on activated carbon.

Lurje[210] has described the separation of activated carbon adsorbate into 8 different fractions. The dried activated carbon is first extracted with ether and then with an azeotropic mixture of 1,2-dichloropropane and methanol. After evaporation of the azeotrope, the latter extract is treated with chloroform and separated into group 1 (chloroform-insoluble) and group 2 (ether-insoluble, chloroform-soluble). After the ether extract is evaporated and treated with a small amount of water, group 3 is obtained (readily water-soluble). The water-insolubles are dissolved in hydrochloric acid. Group 4 (basic compounds) is isolated from this solution with sodium hydroxide and ether. After neutralizing with acetic acid, group 5 (amphoteric compounds) can be isolated with ether. The HCl-insoluble fraction in the form of ether solution is treated with 5% aqueous $NaHCO_3$ solution and NaCl. Stronger acids go into the aqueous solution, are precipitated with HCl and extracted from ether (Group 6). Phenols and other weakly acid compounds (Group 7) are extracted from the fractions of the ether solution which are insoluble in the aqueous $NaHCO_3$ phase by means of 5% aqueous NaOH, while the neutral substances remaining in the ether solution represent group 8.

*Enrichment by evaporation.* In the case of compounds which are not volatile and do not decompose during longer boiling with water, the water samples can be directly evaporated to a small volume. A simple apparatus for unsupervised evaporation to a desired end volume is described on p. 67.

*Isolation of volatiles by outgassing.* Methods to convert volatiles into the gas phase are described on page 98 in connection with head space analysis by gas chromatography.

*Enrichment by freezing.* Pollutants of sufficient water solubility at lower temperatures can be enriched by freezing. The ice which separates after sufficient cooling can be obtained in such compact form that the impurities remain practically completely in the unfrozen liquid and can be decanted. This freezing technique, which has been described for analytical purposes by Baker[16,18] as well as by Kammerer and Lee[157], has the advantage that it prevents losses of volatiles, of pollutants by solvents and decomposition reactions of thermally unstable compounds.

Baker fills a 1-liter round-bottom flask with 200 ml of water sample, places it at a tilt of 60° into a cooling mixture of ice and common salt at −12° and rotates the flask at 80 rpm. After about 20 min the residual liquid of 20-30 ml can be easily decanted from the compact pure ice block; rinsing is not necessary. The method was tested with highly dilute solutions of phenols, volatile fatty acids and acetophenone (0.1-10 mg/l) and proved to furnish yields of more than 90%. The enrichment can be increased by repeating the process (cascade freezing).

*Liquid-liquid extraction.* The following aspects are important in the choice of solvents for liquid-liquid extraction of water pollutants:

1. Favorable position of the distribution equilibrium. The material to be isolated should have optimum solubility in the extractant. A larger quantity of extractant or a greater number of extractions is needed with higher water solubilities. Naturally the extractant itself should be insoluble or only sparingly soluble in water.

2. The extraction can be made selective by adjustment of an appropriate pH. For example, phenols in highly alkaline solution (pH > 12) are retained in the aqueous phase, while they are extractable at pH 8, while more strongly acid compounds are retained in the aqueous phase.

3. The boiling point of the solvent must differ sufficiently from that of the extracted materials if these are to be completely isolated during evaporation.

4. Traces of solvent may not interfere with subsequent procedures. For example, no halogenated hydrocarbons may be used if the components of the sample are to be determined by gas chromatography with the use of an electron capture detector which is particularly sensitive to halogen compounds.

5. The solvents must be of such purity that evaporation for enrichment of the contaminants cannot produce false results.

6. In a subsequent IR- or UV-spectrophotometric analysis, the solvents should be optically transparent in the respective spectral regions.

A thorough distribution of the two liquid phases with the largest possible effective interface during the extraction process is realized either by agitation or by stirring with high-speed blade agitators. The necessary agitating or stirring time is essentially determined by the fineness of solvent droplet division. Agitating or stirring periods of a few minutes up to a few hours are prescribed. On the other hand, excessively energetic agitation or stirring may not lead to the formation of stable emulsions which prevent the subsequent phase separation; in any case, settling periods of several hours must be allowed at times.

Solvents of higher specific gravity than water can be easily collected at the base of conical vessels. Cotton or glass wool plugs placed into the stem of a separatory funnel provide a simple clarification of the solution.

A simple device for the collection and withdrawal of dissolved

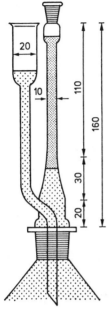

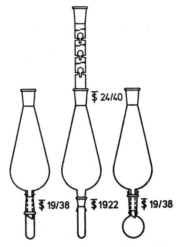

Fig. 3.2a. Microseparator for direct extraction of pesticides from water (units in mm).

Fig. 3.2b. Evaporator according to Kuderna-Danish.

phases of lower density than water has been described by Weil and Quentin[327]. Two-liter flat-front flasks with ground joints serve to shake 1.8-1.9 1 of water sample with 20 ml petroleum ether for 12 h on a shaking machine. After settling for 1 h, and after attachment of a "microseparator" (Fig. 3.2a), water is added to the funnel tube on the left which displaces the petroleum ether layer into the narrow section of the riser on the right and an ali-quot (*e.g.*, 15 ml) is drawn off with a pipet for further treatment.

Several authors have described automatic liquid-liquid extrac-tors (Friedrichs[94], Kahn and Wayman[154], Sanderson and Ceresia[270], and Piorr[250]). The problem is to obtain a small enough droplet size of the extractant to form a sufficiently large phase boundary for rapid adjustment of the distribution equilibrium while at the same time preventing the formation of stable emulsions during the necessary brief phase separation period.

The apparatus of Kuderna-Danish with several exchangeable parts for adjustment to the boiling point and quantity of solution (Fig. 3.2b, see Gunter and Blinn[119]) has been widely recom-mended for the evaporation of larger volumes of solvent from solutions of low extract concentration. For example, 300 ml of a

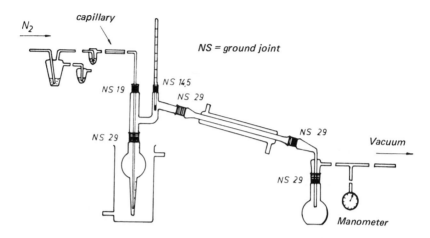

Fig. 3.2c.   Distillation equipment according to Dietz and Koppe.

petroleum ether solution can be concentrated to 2 ml in 10 min without losses.

Ditz and Koppe[63] have described evaporation apparatus for the determination of small quantities of extracts in the $\mu$l-range (Fig. 3.2c). The extract collected in the flask constriction after evaporation can be drawn off with a microsyringe, flushed with minimal quantities of solvent and collected in small graduated flasks of *e.g.*, 100 $\mu$l, from which an aliquot can be removed (for example, by weighing). The sensitivity of techniques requiring small volumes of liquid (*e.g.*, thin-layer or gas chromatography) can thus be considerably increased.

# 4. SUBJECTIVE TESTS

Before the start of an actual analysis and possibly already at the time of sampling, simple subjective tests can reveal important findings without any cost, giving valuable information concerning the quality of a water sample and the presence of contaminants as well as the procedure to be used in analysis. This is a test for odor, taste and possible color of the water sample. A noticeable odor, taste or marked color are sufficient to reject water for human consumption. According to DIN 2000 and 2001, water for drinking and residential purposes must be free from unnatural odors or taste, and colorless or at least free from a marked color.

## 4.1 Odor Test

In order to have a perceptible odor, a waste contaminant must be volatile, *i.e.*, a certain fraction of the sample must be released to the ambient atmosphere in the vapor phase. This fraction will be greater the lower its boiling and water solubility. Odors are produced primarily by organic compounds, while among inorganics, only ammonia and hydrogen sulfide are distinctive. The presence of the latter two compounds as a rule originates from decomposed organic materials, however; mixed odors of these two components occur primarily in stagnant water polluted by excrement or other natural decomposition products. More specific odors from organic matter indicate the presence of industrial wastes.

The detection sensitivity for organic chemicals in aqueous solution by the olfactory organ differs highly, extending over 4-6 orders of magnitude. Moreover, the individual olfactory power of examiners may differ by 2-3 orders of magnitude. In addition, fatigue and adaptation phenomena must be taken into account.

The appendix of ASTM D 1292/65 lists threshold odor values in ppm of aqueous solutions. These are mean values from ratings of several testers. Some of these are reproduced in Table 4.1a.

Odors and flavors are partly of natural origin, for example, sesquiterpenoids, and are produced by algae as well as by some *Streptomyces* species.

Table 4.1a. Threshold odor values in aqueous solutions
in ppm (mean values)

| Formaldehyde | 50 | Pyridine | 0.82 |
|---|---|---|---|
| Acetic acid | 24 | 2,4-dichlorophenol | 0.21 |
| Acrylonitrile | 19 | Octylalcohol | 0.13 |
| Phenol | 6 | Amylacetate | 0.08 |
| n-Butanol | 2.5 | Ethylacrylate | 0.007 |
| | | n-Butylmercaptan | 0.006 |

In practice, odors are usually produced by mixture of substances of varying origins rather than by individual substances, and these then furnish pertinent information concerning water quality. Such mixed odors can be rated as "rotting," "musty," "acidic," etc.

Since the human olfactory sense is little suited for generating quantitative numerical data, a determination of the threshold odor value, i.e., the dilution of a water sample with odor-free water where an odor was barely perceived, must be accepted as sufficient in addition to general characterizations (strong or weak odor). This determination is the subject of DEV B $\frac{1}{2}$ (1971 edition) as well as of the APHA Standard Methods.

*Odor test according to the APHA Standard Methods.* The odor test is made at 40° in odor-free 500-ml Erlenmeyer flasks with ground stoppers. Water with a free chlorine content is pretreated with the precisely necessary quantity of sodium thiosulfate or $Na_2S_2O_3$. For the preparation of odor-free dilution water, a 5-l bottle is filled with pea-sized activated carbon particles and closed with a rubber or cork stopper provided with two bores. An inlet tube leads to the bottom, while the outlet is provided at the neck of the bottle which contains a glass wool plug. The water is passed through at a rate of 100 ml/min. Frequent testing for the absence of odor is necessary.

At least two subjects are needed for the odor test; one subject prepares the mixtures and the other tests the odor. More reliable results can be expected with the use of a larger number of testers. The hands, face and hair of the testers as well as the testing room must be odorless. The sensitivity of the olfactory power can be tested with aqueous solutions of n-butanol (0.05-1 mg/1).

It is advisable to begin with a preliminary test in which 25 ml of the water sample are added to 175 ml dilution water. The mixture is placed into the Erlenmeyer flask which is closed and heated to 40°, is vigorously shaken and after opening of the stopper, is tested for odor. Depending on whether an odor was detected or not, the mixture is diluted or concentrated.

The actual test always begins with an odorless mixture while the concentration is being increased. The dilution which has produced a marked odor in the preliminary test is prepared in several stronger dilutions (2-4 fold) for the actual test. Two flasks with odorless water are kept in readiness for each odor test so that the odor difference can be determined at any time; however, the tester is not informed as to the flasks which contain the blanks.

Threshold odor number (TO) is obtained from the formula

$$TO = \frac{A + B}{A} \quad (A = \text{ml of sample water, } B = \text{ml dilution water}).$$

Thus, for odorless samples, $TO = 1$.

In the case of highly odoriferous waste water, odor intensity index units (OII units) can be listed instead of the TO values. The relationship of the two values results from the formula $TO = 2^{OII}$ or the correlations listed in Table 4.1b.

**Table 4.1b. Dilution schedule and data of results**

| Dilution | ml sample and dilutions A-E brought to 200 ml with dilution water | TO value | OII |
|---|---|---|---|
| Initial sample | 200 | 1 | 0 |
| | 100 | 2 | 1 |
| | 50 | 4 | 2 |
| | 25 | 8 | 3 |
| | 12.5 | 16 | 4 |
| Dilution A | 50 | 32 | 5 |
| (25 ml initial sample | 25 | 64 | 6 |
| brought to 200 ml) | 12.5 | 128 | 7 |
| Dilution B | 50 | 256 | 8 |
| (25 ml dilution A brought | 25 | 512 | 9 |
| to 200 ml dilution) | 12.5 | 1024 | 10 |
| Dilution C | 50 | 2050 | 11 |
| (25 ml dilution B brought | 25 | 4100 | 12 |
| to 200 ml dilution) | 12.5 | 8200 | 13 |
| Dilution D | 50 | 16400 | 14 |
| (25 ml dilution C brought | 25 | 32800 | 15 |
| to 200 ml dilution) | 12.5 | 65500 | 16 |
| Dilution E | 50 | 131000 | 17 |
| (25 ml dilution D brought | 25 | 262000 | 18 |
| to 200 ml dilution) | 12.5 | 534000 | 19 |
| | 6.25 | 1050000 | 20 |

With regard to the theory and practice of odor determinations in water, see also Malz and Gorlas[216] as well as Soucek, Popovska and Sindelar[295].

*Odor test according to DEV, B ½.* The qualitative odor test is made during sampling directly in the sample bottle, in the laboratory in an Erlenmeyer flask covered with a watch glass in which 100 ml of the water sample are heated to about 60°. After swirling the odor is tested and characterized as "earthy, musty, nauseous-reeky, urinous, sewage-like, rotten, cabbage-like, fecal, fish-oil-like, aromatic" or on the basis of certain chemicals (phenol, tars, chlorophenol, volatile acids, aldehydes, resins or mineral oils).

The quantitative odor test according to DEV B ½ (1971 edition) is made on the basis of the APHA specification. It requires a minimum of three examiners. The test is performed at 20°; if no odor is perceived at this temperature, the sample is heated to about 60°. The test begins with 5 formulations: dilution water, 0.1%, 1% and 10% dilution of the water sample in the dilution water as well as the undiluted sample. The formulation from which an odor can be perceived is tested once more in 6 further dilutions.

*Taste test.* A taste test of water samples is advisable only if they are definitely free from pathogenic organisms or toxins. In the case of doubtful samples, the chemical and bacteriological determinations offer sufficient evidence for evaluation. A taste test of drinking water is acceptable if local appearance led to unobjectionable results as well as in chlorinated or ozonated drinking water. The taste test identifies phenols, chlorophenols or mineral oils in the water.

According to DEV, the taste test of drinking water is made at 8-12°; however, a foreign taste becomes more apparent at 30°. The rating is qualitative in words of "insipid, salty, acid, alkaline, bitter, sweetish" or on the basis of certain chemicals.

Odor perception is often superimposed on taste sensations.

## 4.2    Color Determination

In a suitable layer thickness pure water appears to be blue. A yellowish color of water may be due to natural origin, in addition to the presence of Fe(III) and manganese compounds and may be caused by humic acids. The entry of residential or agricultural waste water imparts yellow color to water from urine and fecal components (urochromes and bile pigments). Moreover, numerous industrial effluents, especially pulp wastes, have a yellow to brown color. The yellow color is highly pH-dependent. It is intensified by a higher pH value.

*Color tests.* The DEV test is performed simply in colorless bottles with layer thicknesses of 10-15 cm on a white background. An observed color is characterized as yellowish, brownish-yellow or brown, and the intensity as colorless, very weak, weak or strong.

The standard APHA methods compare the colors in Nessler tubes with a 50 ml mark with standard solutions of 1.246 g $K_2PtCl_6$, 1 g $CoCl_2 \cdot 6H_2O$ and 100 ml conc. HCl diluted to 1 l with water. This solution corresponds to 500 APHA degrees.

The determination can also be made in a Hellige comparator against color disks which are also referred to chloroplatinate solutions.

# 5. QUANTITATIVE EVALUA-TION OF ORGANIC COMPOUNDS

## 5.1 Summary

Apart from the special damage produced by individual water pollutants in terms of a toxic action on man and animals, influence on odor and taste, surface effects, etc., the total organic materials also have an effect on the water. These need not be of unfavorable nature. When correctly metered, many organic materials can serve as nutrients for bacteria and other microorganisms and thus finally form the basis of successful fish hatching. However, adverse effects are much more frequent when the cleanest and most bacteria-free water is desired, for example, for drinking or bathing purposes.

Aerobic (oxygen-requiring) bacteria with a metabolism utilizing organic materials for their nutrition consume oxygen in this process, obtaining it from the reserves dissolved in water. If this consumption is greater than its makeup by fresh dissolved oxygen obtained from the surface, an oxygen deficit is present in the water which in turn can have consequences for the biocenosis existing in the water. More demanding high-quality fish migrate into other zones, while an increasing oxygen deficiency (less than 2-3 mg $O_2$/l) leads to general fish death, and the lower aerobic organisms will gradually also die out. They are replaced by anaerobic bacteria, but with increasing decay, these also die and the water becomes a wasteland.

However, if the load of organic materials and thus the oxygen demand do not exceed a moderate degree and aerobic bacterial life is maintained, the organic pollutants are partly degraded by bacteria by oxidation (dissimilation) and partly used for the growth of new living biomass (assimilation), and in either case they are eliminated from the aqueous solution. Thus, a so-called "self-cleaning effect" has occurred in the water (see 1.2.5) which, however, requires continuous control of the permissible degree of the organic load.

Beyond this, the total organic pollutants can have various unfavorable effects on water. Color and transparency are impaired, undesirable although often nonspecific odor and taste effects become evident, and drinking water treatment with chlorine or ozone requires a higher consumption of oxidants which may be accompanied by further impairment of taste. The food and beverage industries, in particular, require water which is optimally free of organic pollutants, especially since such materials can form the nutrient substrate for secondary infections.

Thus various objectives result for water analysis to determine total organic pollutants: in spring and ground water with a relatively low degree of pollution but even higher purity requirements for drinking water production, in surface water in order to maintain a controlled oxygen balance as well as in view of its use as raw water, if necessary also as drinking water, and finally in more highly polluted residential and domestic waste effluents with a view toward the expected load on the sewer and in the sequence of operations for their purification.

It is apparent from the above that the pollutants which are subject to rapid bacterial degradation and which are a direct load on the oxygen balance of a body of water are of primary interest. They are followed by the remaining organic impurities which are also oxidatively degraded by bacteria in the further course of the process and in any case affect water adversely in some way.

Simple but not very informative methods to determine the total organic pollutants have been known for a long time. More specific analyses became necessary only in recent decades with increasing water pollution. Research for methods and apparatus is still continuing in this field and is the subject of numerous publications.

The oldest method to obtain some insight into the quantity of organic pollutants present is to determine the ignition loss. Ignition of an evaporation residue obtained at 110° reveals many organic substances (carbohydrates, protein compounds) by dark discoloration and carbonization. However, several organics are involved in the ignition loss itself.

The organic carbon determination is also a method in long use. Originally carried out by tedious wet methods (with $KMnO_4$, chromic acid and later also with persulfate) organic carbon has recently been determined by rapid methods with dry combustion in commercial apparatus in which very small volumes of sample (20-50 $\mu$l) are evaporated and combusted. The $CO_2$ formed is determined either by IR-absorption or, after catalytic hydrogenation into methane, by the flame ionization detector (see Page 45). These methods have the advantage of being rapid and quantitative but they only offer information on the carbon of the organic compounds while hydrogen cannot be determined; thus, accurate knowledge on the oxygen demand during oxidation in water, for example, cannot be derived in this way.

Oxidimetric methods in which the consumption of oxidizing titration solution serves as a criterion of the oxidizable pollutant content have also been in use for a long time. In this connection, the determination of the permanganate consumption under precisely set conditions is an old established method of water analysis, even though oxidation by no means takes place quantitatively and many, especially industrial, pollutants are not determined.

Other standard oxidizing solutions, *e.g.*, alkaline sodium hypochlorite solution which reacts primarily with nitrogen-containing pollutants, have a similar action although in a different direction.

A practically complete oxidation effect on nearly all water-soluble and on many undissolved organic pollutants is obtained with standard potassium dichromate solutions in strong sulfuric acid in the presence of silver sulfate as catalyst and mercury(II) as the masking agent of chlorides. Various recently developed improvements led not only to a more rapid procedure but also to a refinement insofar as even very small concentrations of organic pollutants in the range of less than 1 mg/l can be determined.

Even a standard analytical determination with potassium dichromate furnishes a weight indication of the organic pollutants present in the water only if their elemental composition (empirical formula) is known. However, the $K_2Cr_2O_7$ consumption results in the oxygen demand for their complete oxidation into $CO_2$ and $H_2O$. The result is expressed as the COD-value (chemical oxygen demand) (in German known as CSB or chemischer Sauerstoffbedarf).

The COD test can be conducted by various other methods, although the dichromate method is the (APHA) Standard Method in the U.S. Instead of the oxidimetric $CrO_4$ titration, the green Cr(III) ion formed can also be analyzed by spectrophotometry with the use of recording instruments. Persulfate can also serve for a

COD determination in which the water sample is treated with a known quantity of this oxidant and the remaining excess is determined as oxygen by gas volumetry.

Finally, two types of apparatus have recently been described and introduced on the market based on dry microreactions which produce similar COD values very rapidly from small quantities of sample; in the first, carbon monoxide is the reaction product and is determined by IR-spectroscopy, while in the second complete combustion is carried out with a measured quantity of oxygen and the oxygen remaining in the excess is determined by galvanometry.

The methods described thus far all have the purpose of a total organic pollutant determination regardless of whether they are degraded by reactions in natural processes. However, there has long been a need for a method by which only those biodegradable organics are determined which influence the oxygen balance of a waste water or sewer or other body of water in a direct way.

The oldest method of this type—determination of the oxygen demand according to Spitta—involved relatively clean surface waters. The dissolved oxygen content is directly determined in the water sample. Subsequently an identical water sample if allowed to stand for 48 h in the absence of air and the residual oxygen is then determined. The difference in oxygen content yields the "oxygen consumption."

In the following period the method was modified to be applicable to more highly polluted water. To obtain complete conversion the treatment time was extended to 5 days and the increased oxygen demand was covered by dilution with pure air- (or oxygen)-saturated water. The difference between the oxygen present in the initial mixture and that remaining after standing for 5 days referred to 1 l of water sample yields the "biochemical oxygen demand after 5 days" or the $BOD_5$-value.

This "dilution method" has been variously modified. Oxygen also can be introduced in the gas phase and the demand determined by gas volumetry or on the basis of the pressure drop in a closed cycle. The well-known Warburg apparatus as well as a special system developed by Sierp-Fränsemeyer can serve for this purpose. Recently automatic apparatus is being offered in which the oxygen demand is measured by barometry or coulometry.

Very recently the author proposed a simplified method in which the $BOD_5$ value results as the difference of two COD values, one of which is measured immediately and the other after standing for 5 days in the presence of excess air which, however, need not be measured.

An important but hardly avoidable disadvantage of the $BOD_5$ test is that it requires a long time, which can be reduced only conditionally on the basis of special empirical values.

In the evaluation of $BOD_5$-values with consideration of the quantity of organic pollutants present it must be kept in mind, first of all, that only the biodegradable compounds are determined. However, not even these are oxidized quantitatively; a fraction remains unconverted, and another fraction is assimilated into a new biomass. If the $BOD_5$-value is considered as a criterion of the organic pollutants of a body of water, as is generally the custom, these limiting factors must be kept in mind, even though extensive empirical information is available about them, at least for residential waste water.

## 5.2    Determination of the Ignition Loss

Determination of the ignition loss of the evaporation residue of a water sample offers an indication, but not a quantitative statement, concerning the existing organic materials.

*Procedure:* (1) Determination of the evaporation residue: According to the German Standard Methods (DEV), 100 ml of water sample are evaporated in a weighed platinum dish, dried for 1 h at 110° and weighed. If the weight does not deviate by more than 10% after additional drying for 30 min at 110°, it is assumed to be constant. Otherwise, drying must be continued. (2) Ignition residue: The platium dish with the evaporation residue is heated to 600-650° for 15 min and the degree of blackening which appears in the meantime is observed. If dark residues remain, these are wetted with ammonium nitrate solution and heating is then continued to 600-650° for about 10 min. The ignition loss is the difference between evaporation residue and ignition residue in mg/l.

According to the APHA Standard Methods, the evaporation residue is dried at 103° or, especially in the presence of a large amount of organic material, at 179-181° and is then heated to 600° for 1 h to determine the ignition residue.

*Evaluation:* The ignition loss at 600-650° alone by no means indicates the "organic substance" of a water sample. At this temperature, the following inorganic components are at least partly volatilized from the evaporation residue in addition to the combusted organic compounds: water of crystallization and hydration, $CO_2$ from Ca and Mg carbonates, HCl *e.g.*, from $MgCl_2$ by hydrolysis as well as nitrous oxides from nitrate by hydrolysis as well as by reduction. On the other hand, many organic compounds are volatilized already during evaporation. $Na_2CO_3$ forms from the sodium salts of organic acids. The smaller the quantity of organic pollutants, the greater the percentage of the error becomes.

As a rule, therefore, the ignition loss can only be evaluated qualitatively depending on whether the colorless evaporation residue remains colorless during ignition, assumes a pale brownish color or turns an intense black with the formation of tarry fumes.

With a marked predominance of nonvolatile organic compounds, *e.g.*, in sludge samples, the determination of the ignition loss yields results which can be better evaluated quantitatively.

## 5.3 Organic Carbon Determination

### 5.3.1 Summary

The methods for an organic carbon determination in waste water are largely based on experiences in the field of organic elemental analysis. They exist for microquantities (2-20 $\mu$l) as well as for larger volumes of water (10-100 ml). While the workup of larger volumes of water is complicated and time-consuming, the procedure with small samples requires only a few minutes but has the disadvantage that even minor inhomogeneities (sludge particles) cause great deviations in the result. The samples must therefore be available either in the form of clear liquids or in an extremely homogenized state. It is often recommended that several parallel determinations be made and that the "deviants" be screened out according to the rules of statistics when mean values are formed.

Combustion takes place either by the dry method after evaporation of the water in an oxygen stream or by a wet method with a suitable oxidant ($K_2Cr_2O_7$ or $CrO_3$ in sulfuric acid in the presence of $Ag_2SO_4$ or with potassium persulfate $K_2S_2O_8$, also with silver sulfate as catalyst). In all cases, numerous methods are available for the final determination of the formed $CO_2$: gravimetry, acidimetry with visual or potentiometric end-point determination, conductimetry, IR-spectrophotometry, and, after conversion of $CO_2$ into $CH_4$, the flame ionization detector.

Carbon dioxide present in the free state or bound as carbonate, i.e., the "inorganic" carbon, must be eliminated before analysis or determined in a parallel analysis in all cases.

### 5.3.2　Dry Combustion in Larger Volumes of Liquid

Method of Montgomery and Thom[227a]

Principle: Evaporation of up to 50 ml water sample in an $O_2$ stream, ignition of the residue and transport of the vapors over copper oxide at 900°; determination of the formed $CO_2$ by IR-spectrophotometry.

Apparatus: $CO_2$-free oxygen, a quartz flask of 100 ml volume with a fused glass inlet tube, a quartz tube filled with copper oxide which can be heated to 900° in a furnace, a descending condenser, a small wash bottle with acidified $K_2Cr_2O_7$ solution for $SO_2$ retention, a drying tube with Anhydron, a pressure regulator, a spherical gas bottle of brass sheet which can be evacuated and a suitable nondispersive IR-instrument with circulating pump for larger volumes of gas.

Procedure: The water sample (up to 50 ml with not more than 200 μg organic carbon or 330 mg solids) is adjusted to a pH of less than 5.0 with potassium bisulfate solution (15 g/100 ml water) and heated briefly to remove the released $CO_2$. The water is then evaporated and the residue is finally brought to red heat on a Bunsen burner. The combustion gases are transported into the evacuated gas bottle with oxygen and the $CO_2$ content is read in the IR-instrument.

Analysis time: Depending on the solids content, 75 min to 3 h. A C-yield of 100.2 ± 1.9% was obtained with some test substances.

Türkölmez[315] treats 50 ml of water sample by a similar method. Instead of copper oxide he uses cobalt(II-III) oxide at 700° and he determines the formed $CO_2$ by gravimetry in an Ascarite tube. Inorganic $CO_2$ is eliminated by blowing out in an $O_2$ stream at pH 4.6-4.8 (100-120 ml/min oxygen at room temperature for 30-45 min) while any volatile organics are retained in a dry ice trap. After blowing out the $CO_2$ the contents of the cold trap are heated and combusted in an $O_2$ stream. The water sample is then evaporated and the residue is ashed by prolonged ignition with a Bunsen burner, Analysis time: 1.5-3.5 h.

Gorbach and Ehrenberger[112] evaporate and combust 0.1-0.2 ml of the waste water sample in an oxygen stream in a Pt or quartz boat after addition of a few mg $K_2S_2O_7$ and transporting the vapors over copper oxide at 900°. The combustion gases are freed from halogens by means of silver wool and from nitrous oxides by washing with concentrated sulfuric acid. The formed $CO_2$ is determined by potentiometric titration with 0.01 N $NaOH/BaCl_2$ solution in a semi-automatic titration instrument.

Salzer[268] uses a similar method to combust about 0.25-0.5 ml of the waste water sample of 600-800 mg $C/l$ and absorbs the formed $CO_2$ in 0.01 N NaOH in a cell equipped for measurement of the electrical conductivity.

Sander[269] combusts 2-5 ml waste water in a detonating gas flame of the Wickbold quartz apparatus. Halogens are removed with silver nitrate solution, nitrous oxides with conc. $H_2SO_4$. The formed $CO_2$ is absorbed with a solution of $BaCl_2$, NaOH and $H_2O_2$ (pH 11) in a circulation cell and automatically titrated potentiometrically with a mechanical burette. A platinum wire serves as the measuring electrode, while the calomel electrode is the reference. Twelve analyses can be performed per hour.

### 5.3.2.1-2 Dry Combustion of Microvolumes of Water

The method of dry combustion of microvolumes of water with subsequent IR reading of the formed $CO_2$ was developed into a rapid micromethod by van Hall, Safranko and Stenger[121] (see also van Hall, Barth and Stenger[122], van Hall and Stenger[123], Schaffer[271] and 6 co-workers). The apparatus designed by these authors is manufactured and sold by the Beckman Company as the organic carbon analyzer formerly known as the carbonaceous analyzer. One analysis with simultaneous determination of the inorganic ($CO_2$) carbon requires 5 min with a sample volume of 20-40 $\mu$l. The combustion tube is packed with cobalt oxide supported on asbestos and the combustion temperature in an air stream amounts to 950°. The IR-analyzer records the formed $CO_2$ as a peak with a height which represents a measure of the total carbon. A second sample is conducted through a decomposition tube with 85% phosphoric acid supported on quartz chips at 150°, and free and carbonate $CO_2$ but not the $CO_2$ from urea and dicarboxylic acids is liberated without oxidation of the organic matter and is also measured as a peak. The difference in peak heights furnishes the organic carbon value.

The measuring range with full deflection amounts to 50-4000

mg C/l and in a special model to 0-10 mg C/l. The reproducibility is reported to be ±1% of the full deflection corresponding to 0.2 mg C/l. Combustion is quantitative (99-100%) and is not disturbed by secondary impurities.

The reproducibility of the entire determination depends highly on the homogeneity of the sample. In view of the small quantity of sample, inhomogeneities (solid components) cause marked deviations. The most uniform results are therefore obtained with filtered water.

See also Bauer and Schmitz[20] concerning practical experiences with this method.

A continuously recording instrument based on the same principle has been described by Axt[14]. The water sample is continuously evaporated with the necessary quantity of air for combustion and combusted over quartz wool at 600°. The $CO_2$ formed by combustion is continuously measured by IR absorption; carbonate $CO_2$, after acidifying with nitric acid, is outgassed previously in a tube packed with glass splinters. Compared to a single determination, the measuring accuracy is increased; for example, 0.45 ± 0.03 mg org. C was detected in the Karlsruhe tap water. The apparatus can also be used as a detector for water-soluble organic components in column chromatography of aqueous solutions (see 6.3).

### 5.3.2.3. Methane Method according to Cropper, Heinekey and Westwall[59]

*Principle.* The sample is combusted, the formed $CO_2$ is hydrogenated with hydrogen to form $CH_4$ and the latter is indicated by a flame ionization detector.

*Procedure.* A sample of 2 μl with 1-10 μg C is evaporated in a nitrogen stream and combusted over copper oxide at 850-900°. Hydrogen is added to the combustion gases and the mixture is hydrogenated at 300-350° over a nickel catalyst on ground brick. Subsequently, the product is dehydrated on silica gel and combusted in a flame ionization detector with added oxygen. The peak area recorded on the recorder serves as a measure of the C-content of the sample. Calibration is made with methane or with a suitably diluted potassium biphthalate solution. Carbonate $CO_2$ is determined in parallel by injection into a tube heated to 150°; the difference furnishes the organic carbon. The reproducibility is reported to be ±5% of the carbon content; the analysis time

amounts to 2 min. The same authors also report on a further en-
hancement of sensitivity to 1 ppm C in the water sample or 80%
scale deflection by 0.08 $\mu$g C (see also Dobbs, Wise and Dean[66]).

### 5.3.3  *Determination of Organic Carbon by Wet Oxidation*

#### 5.3.3.1  Procedure According to the German Standard Methods (DEV) First Edition, with Chromic Acid

*Apparatus.* A wash bottle containing 33% potassium hydroxide
solution and a soda lime tube for purification of the air serving
as the carrier gas, a boiling flask with dropping funnel and reflux
condenser, a wash bottle with 10% potassium iodide solution for
absorption of elemental chlorine from chlorides, another wash
bottle with 20% sodium thiosulfate solution, a bulb with concentrated sulfuric acid to remove steam, and a weighed potash
bulb containing 50% potassium hydroxide solution.

*Procedure.* The flask is filled with 10-40 mg C-containing water
sample, previously acidified and freed from $CO_2$ by gentle boiling in an air stream, 50 ml chromic acid mixture, 50 ml sulfuric
acid-phosphoric acid mixture and 2 ml concentrated sulfuric
acid per ml sample. Heating is carried out for 2 h up to gentle
boiling while $CO_2$-free air is bubbled through (100 bubbles/min).
The $CO_2$ formed by oxidation is weighed in the potash bulb.
Chromic acid mixture: 340 g $CrO_3$ are dissolved in 400 ml hot
$CO_2$-free water and brought to 1 l with 85% phosphoric acid.
Sulfuric-phosphoric acid mixture: 1 vol. conc. $H_2SO_4$ + 1 vol.
85% $H_3PO_4$.

Egli-Schär[73] boils water samples which proved to be difficult to
decompose during treatment with $K_2S_2O_8$ (see below), with a
triple volume of concentrated sulfuric acid and 5 g $K_2Cr_2O_7$ in
the presence of 1 g silver sulfate for about 30-45 min in a nitrogen
stream. The $CO_2$ formed is absorbed and titrated as described on
p. 81.

According to earlier experiences with wet carbon analysis with
$CrO_3$ in concentrated sulfuric acid, the losses due to formation of
CO or $CH_4$ are small. These substances can be completely combusted by post-combustion of the oxidation gases in a quartz tube
packed with copper oxide or cobalt oxide on asbestos and heated
to 750-900°.

With a water sample of up to 100 ml, Pickhardt, Oemler and
Mitchell[248] use approximately double the volume of concentrated

$H_2SO_4$ and 10-15 ml 50% aqueous $CrO_3$ solution. For post-combustion the oxidation gases are conducted through a quartz tube packed with CuO and heated to 750°. The boiling time amounts to 10 min.

Maier[215] reports on wet carbon analyses with $K_2Cr_2O_7$ and $Ag_2SO_4$ in concentrated $H_8SO_4$ and by post-combustion on cobalt oxide on asbestos. The formed carbon dioxide is measured in an IR-gas analyzer.

The simultaneous determination of organic carbon and the COD value with $K_2Cr_2O_7$ is described on p. 61.

Apparatus for a continuous carbon analysis by thermal conductivity measurements has been described by Kieselbach[164]. The sample is continuously precipitated with $Ba(OH)_2$, filtered, mixed with 0.1 M $CrO_3$ solution in 95% sulfuric acid in a quantitative ratio of 1 vol. water sample + 10 vol. oxidant and heated to 250° for $2\frac{1}{2}$ min. The formed $CO_2$ is driven off in an oxygen stream (10 ml/min), any formed gaseous chlorine is absorbed with metallic antimony, and water vapor is removed with Drierite.™ Finally the $CO_2$-$O_2$ mixture is conducted into the measuring section of a Gow-Mac thermal conductivity cell, $CO_2$ is absorbed with Ascarite and the residual carrier gas is conducted into the reference section of the cell. A scale expansion from 50 to 1000 ppm is possible.

### 5.3.3.2 Wet Oxidation with $K_2S_2O_8$ (potassium persulfate) according to Gertner and Ivecovich[102]

For apparatus, see Fig. 5.3.3.2. A 1-liter three-neck flask with ground joints and condenser, 4 U-tubes packed with pumice-conc. $H_2SO_4$ (E), calcium chloride (F), soda asbestos (Ascarite) (G) and half with soda asbestos and half with $CaCl_2$ (H).

*Procedure.* The water sample of 200 ml is treated with 1 ml dilute $H_2SO_4$ (1 + 3 vol.) To blow out the liberated $CO_2$, a strong $CO_2$-free carrier gas stream is passed through. The U-tubes are then connected, 10 ml 10% $AgNO_3$ solution, 50 ml dil. $H_2SO_4$ (1 + 3 vol.) and 100 ml saturated $K_2S_2O_8$ solution are added through the dropping funnel. The mixture is heated to 80° and held at this temperature for 10 min. This is followed by an addition of 100 ml $K_2S_2O_8$ solution, followed by holding for 5 min at 80°. The mixture is then heated to boiling and the formed $CO_2$ is driven by a gas stream of 4-5 bubbles/sec into the two weighed U-tubes G and H for complete absorption.

In test solutions with oxalic acid, urea, hippuric acid, cystine, dextrose, acetanilide and uric acid, 96-107% of theory were recovered.

In a similar manner, Egli-Schär[73] operates with 100 ml water sample and 20 g $K_2S_2O_8$ and absorbs the formed $CO_2$ in two receivers with 100 ml 0.2 N KOH and 50 ml 0.2 N $Ba(OH)_2$. The absorbed carbonate is precipitated from the NaOH charge with barium chloride and is titrated with 0.2 N HCl against phenolphthalein.

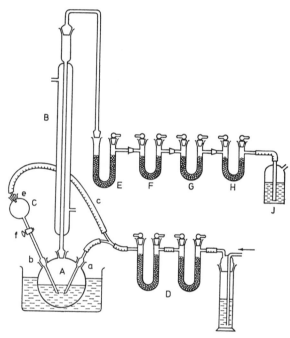

Fig. 5.3.3.2. Apparatus for organic carbon determination.

Leibnitz, Behrens, Koll and Richter[184] use a semimicromethod of Katz, Abraham and Baker[158] for industrial waste effluents: a 50 ml Erlenmeyer flask with an internally fused glass vesesl of 12 mm diameter and 30 cm height with a lateral opening closed with a rubber membrane.

*Procedure:* The water sample of 10 ml in the flask is treated with 0.5-0.6 g $K_2S_2O_8$, a few drops of dilute $H_2SO_4$ and 1 ml 4% $AgNO_3$ solution. The inside beaker is charged with exactly 1 ml 0.5 N NaOH. An injection needle is inserted through the rubber membrane cap, the flask is evacuated, the needle is withdrawn, and the flask is immersed in a water bath of 40-50°, heated for 15-20 min to 70-75° and held at this temperature for 30 min. After cooling, the vacuum is released by piercing the membrane, the rubber plug is removed, the sodium hydroxide solution is siphoned into a titration vessel, precipitated with a few drops of saturated $BaCl_2$ solution, and titrated with 0.1 N HCl against phenolphthalein.

Katz et al.[158] found $CO_2$ yields of 96-102% in 16 different organic substances (alcohols, carbohydrates, organic acids, amino acids).

*Wet combustion according to Abrahamczik, Groh, Huber and Kraus*[5]. Determination of organic carbon by wet combustion in an oxygen stream, with $CrO_3$ + $KIO_3$ or with $K_2S_2O_8$ as the oxidant. The carbonate carbon is acidified and then blown out with phosphoric acid, while condensables are frozen out in a cold trap. Oxidation is completed in a combustion tube connected to the cold trap, packed with silver permanganate which has been decomposed by calcining and heated to 550°; the condensate from the cold trap is also vaporized into this combustion tube. The formed $CO_2$ is collected in a receiver containing dimethylformamide with 0.7% monoethanolamine and is titrated in the absence of water and atmospheric $CO_2$ with 0.05 N tributylmethylammonium hydroxide solution against thymol phenolphthalein.

The oxidation solution consists either of a solution of 20 g $CrO_3$ + 2 g $KIO_3$ in a mixture of 170 ml conc. $H_2SO_4$, 30 ml 85% phosphoric acid and 40 g $P_2O_5$, or a mixture of 15 ml 10% $AgNO_3$ solution and 100 ml saturated $K_2S_2O_8$ solution. The $CO_2$ end determination can also be made by conductivity measurement according to Malissa[215a] in the Westhoff apparatus.

Opperskalski and Siebert[243] carry out the above oxidation processes fully automatically in batches with suitable apparatus. The formed $CO_2$ is measured and recorded in IR apparatus. Instrumental details can be learned from the cited publication.

### 5.3.4    Evaluation of Water Samples Based on the Organic Carbon Content

As mentioned, an organic carbon determination of water samples furnishes a rapid and reliable measure of the sum of organic pollutants of the water. However, a precise quantitative value of the respective substances cannot be obtained without a knowledge of their elemental composition. Nor does it offer exact information on the expected oxygen demand for complete oxidation of these components in the sewer or a comparison with the $BOD_5$ value, since this also requires a knowledge of oxidizable hydrogen and of the already existing oxygen in the molecules.

As indicated by Table 5.3.4, 1 mg organic carbon in oxalic acid corresponds to 3.75 mg organic matter and to a COD of 0.67 mg (for complete oxidation); 1 mg C in phenol in contrast corresponds to only 1.31 mg substance but to 3.22 mg in COD. Thus notable differences in these conversion factors may be expected in industrial effluents.

In residential waste water and industrial effluents similar to it, *e.g.*, from the food industry, in which the oxidizables consist mainly of carbohydrates and protein degradation products, an approximate average factor of 2.4-2.5 mg substance or 2.7 mg COS per mg organic C/l may be assumed. If we now postulate a 70% degradation in the $BOD_5$ determination of such readily biodegradable materials (see p. 72), then 1 mg organic C/l offers a guideline value of about 2 mg/l $BOD_5$ (see also p. 84), which decreases, however, when more sparingly biodegradable components are present, for example, from the effluents of biological treatment plants.

Table 5.3.4. Relation between mg org. C, mg org. matter and mg COD/l

| 1 mg org. C in | mg org. Substance/l | mg COD/l |
|---|---|---|
| Oxalic acid | 3.75 | 0.67 |
| HCN | 2.25 | 1.33 |
| Glycocoll | 3.13 | 2.0 |
| Alanine | 2.58 | 2.67 |
| Glucose. lactic acid, acetic acid | 2.50 | 2.67 |
| Saccharose | 2.38 | 2.67 |
| Proteins | 2.0 | 2.8 |
| Sulfite waste (dry org. matter) | 2.0 | 3.2 |
| Butyric acid | 1.83 | 3.33 |
| Ethanol | 1.92 | 4.0 |
| Butanol | 1.54 | 4.0 |
| Phenol | 1.31 | 3.22 |
| Dodecylbenzene sodium sulfonate | 1.61 | 3.78 |

## 5.4    Methods of Volumetric Analysis

### 5.4.1    Permanganate Methods

The oxidation of water samples with $KMnO_4$ in acid solution dates back to Kubel and that in alkaline solution to Schulze. Holluta and Hochmüller[139] have described a combination of the two types of reactions.

According to the German Standard Methods (DEV), the Schulze method should be used when the chloride ion concentration is greater than 300 mg/l since oxidation into free chlorine would then occur in acid solution.

The Kubel method was first used primarily to test drinking water and surface waters with a low pollution load. The "permanganate consumption" became one of the most important conventional values to characterize organic pollutants, although no stoichiometric reference was made to their nature and quantity. More highly polluted water was tested by this method only at a later time. To avoid the need to dilute these samples excessively with pure water, the use of 0.1 N $KMnO_4$ solution was recommended; however, the effect of $KMnO_4$ depends highly on concentration. A higher degree of oxidation is produced by 0.1 N solutions in undiluted waste water than when the 10-fold dilution is treated with 0.01 N $KMnO_4$ solution.

### 5.4.1.1 Oxidizability with $KMnO_4$ in Acid Solution (Permanganate Consumption) according to DEV, H 4

*Procedure.* A water sample of 100 ml with 5 ml 25 vol.% $H_2SO_4$ in a 300 ml Erlenmeyer flask with attached condenser bulb is heated to boiling, treated with 15.0 ml 0.01 in N $KMnO_4$ solution and maintained at uniform gentle boiling for exactly 10 min. Subsequently 15.0 ml 0.01 N oxalic acid solution is added. The hot solution, which has become colorless, is back-titrated with 0.01 N $KMnO_4$ solution until a barely visible pink color is stable for at least 30 sec. The consumption should amount to between 5 and 12 ml 0.01 N $KMnO_4$ solution. More highly polluted water is suitably diluted with water which is stable to $KMnO_4$ or which has been acidified with $H_2SO_4$ at boiling temperature and treated dropwise with 0.01 N $KMnO_4$ solution up to a stable pale pink color.

Calculation:

$$\text{mg } KMnO_4/l = \frac{[(15 + a) \cdot f - 15] \cdot 0.316 \cdot 1000}{b}$$

a = consumption of 0.01 N $KMnO_4$ in ml,
b = water volume used in ml

$$f = \text{factor of 0.01 N } KMnO_4 \text{ solution} = \frac{15}{\text{ml consumption}}$$

To determine the factor of the $KMnO_4$ solution, a well-titrated hot mixture is treated with 15.0 ml 0.01 N oxalic acid solution and titrated with $KMnO_4$ solution up to a stable pink color. The consumption should be between 14.5 and 15.5 ml $KMnO_4$ solution.

According to the effective DEV specification, boiling is carried out without an anti-bumping plate where occasional bumping and superheating due to nucleate boiling cannot be avoided. In earlier editions, the addition of a pinch of calcined pumice was recommended.

In England and the USA, the $KMnO_4$ consumption in acid solution is determined after standing for 4 h at 27°.

*Oxidizability in alkaline solution according to DEV.* A water sample of 100 ml is treated with 0.5 ml NaOH solution (330 g NaOH dissolved in 670 ml water) and heated to boiling; 15.0 ml 0.01 N $KMnO_4$ solution are added to the boiling solution and maintained at boiling for 10 min with the attached bulb condenser, the solution is then treated with 5 ml 25 vol.% sulfuric acid and 15.0 ml 0.01 N oxalic acid, and worked up as described above.

## 5.4.1.2 Combined $KMnO_4$ Oxidation in Alkaline and Acid Solution according to Holluta and Hochmuller[139]

A water sample of 100 ml in a 250 ml two-neck round-bottom flask with reflux condenser and tubular immersion heater is brought to boiling with 5 ml 15% NaOH solution and 15.0 ml 0.01 N $KMnO_4$ solution on an electric hot plate for 7 min and is then gently boiled for 16 min. Subsequently 7 ml $H_2SO_4$ (1 + 1 vol.) are carefully added through the sidearm, boiling is continued for 10 min, 15.0 ml 0.01 N oxalic acid solution are added, the excess is back-titrated with 0.01 N $KMnO_4$ solution up to a barely visible pink. With pure water, blank values of 0.75-0.95 ml 0.01 N $KMnO_4$ are obtained.

Holluta and Hochmüller performed test analyses with 71 different pure substances using the normal $KMnO_4$ method in acid solution, their combined alkaline-acid $KMnO_4$ method as well as $K_2Cr_2O_7$ in 1+1 sulfuric acid in the presence of silver sulfate (see p. 59). The weakest oxidation effect was obtained with $KMnO_4$ in acid solution. Numerous substances, such as ethanol, glycol, glycerol, acetone, fatty acids, dicarboxylic acids, amino acids, benzoic acid, phthalic acid and dioxane, remained practically unattacked. The highest oxidation values of more than 80% were obtained from phenols. Sugars showed about 30-40% oxidation. Only the method with $K_2Cr_2O_7$ in the presence of $Ag_2SO_4$ gave oxidation yields of 95-98% in nearly all cases (see p. 58). With waste effluents from numerous different industries, the combined alkaline-acid $KMnO_4$-oxidation resulted in approximately double the oxygen consumption compared to the acid $KMnO_4$ oxidation but still remained 20-70% behind the $K_2Cr_2O_7$ oxidation.

### 5.4.1.3 Determination of the Oxygen Consumption with $KMnO_4$

In order to determine the oxygen consumption of the water sample during oxidation with $KMnO_4$, the factor 0.08 ($= 0/200$) is used instead of 0.316 ($= KMnO_4/500$). Moreover, however, the blank $KMnO_4$ consumption must be subtracted which, as an empirical-conventional coefficient, is neglected in the calculation of the permanganate consumption (Leithe[199]). It is determined with reliably pure water, most suitably water which was distilled from a $KMnO_4$ and $H_2SO_4$ solution. If necessary, the water is redistilled over a small amount of NaOH.

The permanganate consumption of pure water including the excess needed to obtain the pink color amounts to about 0.5 ml 0.01 N $KMnO_4$.

The oxygen consumption is calculated according to the formula:

$$\text{mg oxygen} = \frac{(a-b) \cdot f \cdot 8000}{c}$$

a $=$ ml $KMnO_4$ in the sample
b $=$ ml $KMnO_4$ in the blank test
f $=$ normality of the $KMnO_4$ solution (*e.g.*, 0.01)
c $=$ ml of water sample

### 5.4.1.4 Degree of Oxidation Effect of $KMnO_4$ in Residential and Industrial Waste Effluents

An estimate of the oxidation effect of $KMnO_4$ on residential waste effluents can be made on the basis of the biochemical oxygen demand. According to DEV the $BOD_5$ value amounts to approximately 0.67-0.83 times the $KMnO_4$ consumption in mg $KMnO_4/l$ (Table 5.11.1.1). Since the $BOD_5$ value is known to amount to about 70% of the oxygen required for complete oxidation of the organic matter in the water sample into $CO_2$ and $H_2O$, the oxygen consumption with $KMnO_4$ amounts to about 25% of the oxygen demand required for this complete oxidation as obtainable with $K_2Cr_2O_7$ according to the method described on p. 60.

For residential waste effluents, this percentage is surprisingly constant and has recently been confirmed again by Leschber and Niemitz[200a].

However, such a constancy cannot be expected in industrial waste effluents. As indicated by the data of Holluta and Hochmüller, numerous organic compounds, such as alcohols, ketones, fatty acids and amino acids are attacked little if at all, while others, such as phenols or maleic acid, for example, are oxidized almost completely into $CO_2$ and $H_2O$. In the presence of such pollutant mixtures it is therefore impossible to draw even a fairly valid conclusion concerning the organic pollutant load on the basis of the permanganate consumption.

### 5.4.1.5  Evaluation of Water Samples on the Basis of the Permanganate Consumption

The $KMnO_4$ consumption (mg $KMnO_4$/l water sample) determined according to the DEV method is customarily used as a characteristic value of the degree of purity of a water sample, especially of drinking water without consideration of the completeness and nature of the oxidation reaction taking place. In this regard, pure spring and ground water shows $KMnO_4$ consumption values of 3-8 mg/l. Values of more than 12 mg/l are subject of concern for drinking water. Pure surface waters (saprobic stage I-II) show values of 8-12 mg, moderately polluted rivers (saprobic stage II-III) 25-30 mg, highly polluted rivers (stage III-IV) 100-150 mg and more. Swamp water has an elevated $KMnO_4$ consumption of more than 80 mg $KMnO_4$/l.

In the evaluation of the $KMnO_4$ consumption of pure water samples, it must be kept in mind that this parameter according to the DEV standard also contains the blank value formed by self-decomposition of $KMnO_4$ under the experimental conditions as well as the quantity of $KMnO_4$ required for visualization of the pink color. Depending on the procedure used (degree of heating, presence of anti-bumping aids) this value differs, although according to Leithe[199] it may be assumed to be about 1.5-2 mg $KMnO_4$/l.

A study by Leithe[199] demonstrates how completely the $KMnO_4$

consumption reflects the organic pollutants in pure drinking and surface waters. If the oxygen consumption in the form of $KMnO_4$ is compared with the chemical oxygen demand (COD value with $K_2Cr_2O_7$) (see p. 60), we find that only about 25% of the oxygen required for complete oxidation is consumed with $KMnO_4$. For an evaluation of the quantity of organic pollutants in a water sample, a determination of the COD value with $K_2Cr_2O_7$, which leads to practically complete oxidation of the organic pollutants, is therefore considerably more suitable than the $KMnO_4$ test, even though a determination of the $KMnO_4$ consumption in its customary form does furnish useable results because of the large volume of empirical data available, for example, in "small-scale drinking water analysis."

## 5.5    Determination of the Chemical Oxygen Demand (COD Value) with $K_2Cr_2O_7$

### 5.5.1    Summary

The use of potassium dichromate in strong sulfuric acid solution to characterize the organic matter load in water was proposed for the first time by Moore, Kroner and Ruchhoft[230]. Muers[233a] as well as Moore, Ludzak and Ruchhoft[231] suggested the addition of silver sulfate as catalyst to complete the oxidation effect in substances which are otherwise difficult to oxidize, such as acetic acid and amino acids, while Dobbs and Williams[65] recognized that the addition of mercury(II) sulfate was a successful agent to avoid the formation of elemental chlorine from the chlorides with an additional consumption of chromate. It became customary to calculate the chromate consumption as the COD-value (chemical oxygen demand) which resulted in mg oxygen by back-titration of the excess of chromate utilized with an adjusted Fe(II) solution against ferroin as indicator. In this form the method was accepted by APHA in the 12th edition of "standard metohds," and is included also in the 13th edition.

In testing the oxidation action of $K_2Cr_2O_7$ in the presence of $AgSO_4$ with the most diverse model substances and waste effluents, several authors (Moore, Kroner and Ruchhoft[230], Moore, Ludzak and Ruchhoft[231], Holluta and Hochmüller[139], Dobbs and Williams[65], Leithe[189]) found that nearly all tested substances are at least 95-98% degraded into $CO_2$ and $H_2O$, while the ammonia, as well as the amino-, amido- and nitrile-nitrogen is converted into ammonium sulfate with no oxygen consumption. The error of 2-5% can be explained by volatile oxidation-resistant decomposition products ($CO$, $CH_4$) rather than by fractions of the initial materials which remained unoxidized. Some insolubles present in waste water, such as pulp and wool fibers, yeast suspensions and waste-water sludge, can also be almost completely oxidized.

An oxidizability of 95-98% has been reported for the following compounds: methanol, ethanol, butanol, glycol, glycerol, mannite, formaldehyde, acetaldehyde, acetone, formic acid, acetic acid, butyric acid, isobutyric acid, soap, oxalic acid, adipic acid, succinic acid, maleic acid, lactic acid, tartaric acid, citric acid, pyromucic acid, glucose, saccharose, lactose, sorbose, glycocoll, aminocaproic acid, valine, glutamic acid, cystine, histidine, formamide, acetamide, peptone, albumin, casein, KCN, $K_3Fe(CN)_6$, KCNS, phenol, o-cresol, $a$-naphthol, pyrocatechol, pyrogallol, p-benzoquinone, p-chlorophenol, 8-oxyquinoline, nitrobenzene, benzoic acid, salicylic acid, phthalic acid, p-aminobenzoic acid, sulfanilic acid, phenylacetic acid, cinnamic acid, aniline, benzidine, $a$- and $\beta$-naphthylamine, piperidine, toluenesulfonic acid, aminonaphthol disulfonic acid, dodecylbenzenesulfonic acid, dioxane, tetrahydrofuran, indole, 3-methylindole, protein hydrolysates.

Only pyridine and a few other N-containing heterocyclic compounds (pyrrole, pyrrolidine, proline, nicotinic acid) as well as sparingly water-soluble hydrocarbons, such as benzene and homologs, paraffins and naphthenes, have proved to be unoxidizable with $K_2Cr_2O_7$ under the experimental conditions utilized.

The result of practically complete oxidizability of nearly all materials to be expected in dissolved form in residential and industrial waste effluents makes it justified to consider the COD value with $K_2Cr_2O_7$ to be equivalent to the oxygen demand required for complete oxidation (TOD = total oxygen demand) and to use this definition also for residential and industrial waste effluents (see also Leithe[190,191,198]).

In the older literature, the term "chemical oxygen demand" is occasionally used for the oxygen consumed in about 2 hours for rapid nonbacterial oxidation of primarily inorganic components (Fe(II), $SO_2$, $H_2S$) in water in the presence of air, in contrast to the "biochemical oxygen demand" primarily of bacteria which become evident only after a long time (see p. 75).

Since this concept is more difficult to define analytically and the corresponding individual inorganic compounds can be determined more easily, the term "chemical oxygen demand" should be used in the future only in the context of complete oxidation of the organic matter.

Other methodological studies for the determination of the chemical oxygen demand with $K_2Cr_2O_7$ first led to a considerable reduction of the reaction period from the previous 2 h to 10 min (Leithe[192]); furthermore, the sensitivity was increased with the use of relatively pure drinking and surface water, in which even mg-fractions of the COD values/l can now be determined reliably (Leithe[193,199]). (It is still 2 hours in U.S. Standard Methods.) Finally methods were developed for a combination of the COD determination with that of organic carbon and nitrogen (Leithe[189]).

## 5.5.2    Procedure of the Standard Methods (APHA)

A water sample of 20 ml and 10.0 ml 0.25 N $K_2Cr_2O_7$ solution are added to 0.4 g $HgSO_4$ in a 250 ml ground-joint flask with reflux condenser. Concentrated sulfuric acid (30 ml) containing $Ag_2SO_4$ is added in small portions with thorough swirling as well as a few glass beads as anti-bumping aids, and boiling is carried out for 2 h with refluxing. The mixture is diluted to about 140 ml with distilled water, treated with 2-3 drops ferroin indicator and titrated with 0.1 N $Fe(NH_4)_2(SO_4)_2$ solution for a few seconds up to a stable color change from blue-green to reddish-brown. The blank value is determined with 20 ml distilled water under the same conditions.

With the use of 50.0 ml sample, 25.0 ml $K_2Cr_2O_7$ solution, 75 ml $Ag_2SO_4$-containing concentrated sulfuric acid and 1 g $HgSO_4$ are used in the flask, followed by boiling for 2 h, dilution to 350 ml and titration with 0.25 N Fe(II) solution. Reagents: conc. $H_2SO_4$, anal. grade, containing 10 g dissolved $Ag_2SO_4$. Ultrapure $HgSO_4$. An addition of 0.4 g per 20 ml water sample (or 1 g/50 ml) suffices to mask 40 (100) mg Cl or 2000 mg/l. See Cripps and Jenkins[58] as well as Burns and Marshall[46] concerning measures to be taken with higher chlorine contents.

0.25 N $K_2Cr_2O_7$ solution: 12.26 g $K_2Cr_2O_7$, anal. gr., dried at 103°, per liter. 0.1 N $Fe(NO_4)_2(SO_4)_2$-solution: 39 g $Fe(NH_4)_2$ $(SO_4)_2 \cdot 6H_2O$ per liter after addition of 20 ml conc. $H_2SO_4$. The solution must be standardized daily. For this purpose, 10.0 ml of the 0.25 N $K_2Cr_2O_7$ solution are diluted to 100 ml, treated with 30 ml conc. $H_2SO_4$, cooled and titrated after addition of 2-3 drops ferroin. The normality of the solution results from the formula:

$$\frac{ml\ K_2Cr_2O_7 \cdot 0.25}{ml\ Fe(NH_4)_2(SO_4)_2} = c$$

Calculation:

$$mg\ COD/l = \frac{(a-b) \cdot c \cdot 8000}{ml}$$

a = ml $Fe(NH_4)_2(SO_4)_2$ in the blank value
b = ml $Fe(NH_4)_2(SO_4)_2$ to sample
c = normality of the $Fe(NH_4)_2(SO_4)_2$ solution

The reproducibility is reported as a standard deviation of $\pm$ 0.07 ml titration solution with 50 ml of sample.

### 5.5.3 Rapid Method for COD Values of 100-700 mg/l

One disadvantage of the Standard Methods procedure is the considerable amount of time required because of the 2 h boiling period. More recent studies of Leithe[192] show that the boiling time can be reduced to 10 min if the sulfuric acid concentration is increased from a volume ratio of 1:1 to 1.33:1 $H_2SO_4$:aqueous solution.

*Procedure:* 0.25 N $K_2Cr_2O_7$ solution (10.0 ml) is pipetted to 20.0 ml water sample in a 250 ml ground-joint round-bottom flask, 5 small glass beads as an anti-bumping aid (about 3 mm diameter) and 0.4 g $HgSO_4$ are added, followed by 40 ml $Ag_2SO_4$-containing concentrated sulfuric acid (anal. gr) in small portions with thorough swirling; after addition of the first portion, shaking continues until the $HgSO_4$ is completely dissolved. The reflux condenser is attached on the ground joint and the system is heated with a small, nearly non-luminous open gas flame with chimney without use of an asbestos wire screen until gentle but steady boiling is obtained which is maintained for 10 min (about 1 drop condensate reflux in 2 sec). The system is then cooled externally by immersion in cold water, about 50 ml pure chromic acid-treated distilled water is flushed through the condenser, cooling is completed under running water, 2 drops M/40 ferroin solution are added and the product is titrated with $Fe(NH_4)_2(SO_4)_2$-solution which is somewhat stronger than 0.1 N from a 25 ml buret until the color change is obtained from yellow-green via blue-green to reddish-brown. The end point can be clearly identified within one drop.

To determine the normality of the Fe(II) solution, 10.0 ml 0.25 N $K_2Cr_2O_7$ solution are again pipetted into a sharply titrated mixture and titration is carried out with the Fe(II) solution up to the same end point.

The reagents are prepared and the result calculated by the above procedure of the standard methods. The somewhat stronger than 0.1 N Fe(II) solution contains 40-42 g $Fe(NH_4)_2(SO_4)_2 \cdot 6H_2O$ and 20 ml conc. $H_2SO_4$ per liter.

The blank value is determined under exactly identical conditions with 20 ml ultrapure water which is treated with 1:1 chromate-sulfuric acid.

### 5.5.3.1  Procedure for COD Values of 1-50 mg/l
### For Drinking and Surface Water (Leithe[193,199])

A water sample of 50 ml, 10.0 ml 0.05 N $K_2Cr_2O_7$-solution, 1 g $HgSO_4$ and 80 ml $Ag_2SO_4$-containing conc. $H_2SO_4$ are used; otherwise, the method for 100-700 mg/l is followed. After 10 min boiling (where gentle uniform boiling is particularly important), the system is well cooled, about 80 ml $K_2Cr_2O_7$ treated water are added through the condenser, followed by complete cooling, addition of 2 drops ferroin and titration with 0.025 N $Fe(NH_4)_2(SO_4)_2$-solution. Again, the end point can be observed within one drop if some care is used.

0.05 N $K_2Cr_2O_7$-solution contains 2.452 g/l and the 0.025 N Fe(II)-solution 10 g $Fe(NH_4)_2(SO_4)_2 \cdot 6H_2O$ and 20 ml conc. $H_2SO_4/l$.

The blank value is determined with 50 ml ultrapure distilled water which is resistant to $K_2Cr_2O_7$. If necessary, the water has to be obtained by distillation from chromate-sulfuric acid of 50 vol.% $H_2SO_4$. This stability is determined by concentrating 250 ml of the water to 50 ml and repeating the determination. The blank value of the concentrated sample should not be significantly increased.

The standard deviation of a fairly large number of blank tests was found to be ±0.1 ml 0.025 N Fe(II)-solution corresponding to a limit of detection of ±1 mg COD/l with a 95% confidence level.

Further reduction of the limit of detection by evaporation: In the case of particularly clean water, e.g., for drinking water production, a further decrease of the limit of detection of organic pollutants may become desirable so that even mg-fractions of the COD values can be determined reliably. Since such organic materials, which are primarily of natural origin, are nonvolatile, the sensitivity of the method can be increased to any degree by evaporation of the water sample.

For example, 250 ml of the water sample are evaporated to 50 ml in a 250 ml round-bottom flask which has been weighed with 5 glass beads and 3-5 drops 25 vol.% $H_2SO_4$. The process can be carried out without supervision by mounting the flask on one end of a balance-beam device and applying a load on the other end so that after a 50 ml volume of evaporated solution has been reached, the flask is lifted and thus removed from the heating source.

By evaporating, for example, to $1/5$ of the original volume, it is possible to determine whether the water used for the blank value determination, preparation of the 0.05 N $CrO_4^{2-}$ solution as well as for dilution purposes is sufficiently stable to $CrO_4^{2-}$.

### 5.5.3.2 Simultaneous Determination of the COD Value and Organic Carbon Content (Leithe[189])

It seems indicated to combine the determination of the COD value and of organic carbon since it may be expected that the carbon in the practically quantitative oxidation of organic matter is just as completely converted into carbon dioxide. In order to trap and titrate the latter, a wash bottle with a frit (see Fig. 5.5.3.2) containing 10.0 or 20.0 ml adjusted (about 0.2 N) $Ba(OH)_2$ solution treated with 3 drops of n-butylalcohol, is connected downstream of the reflux condenser. According to previous experience (Leithe[186]) this addition of butyl alcohol leads to the formation of a small-cell foam which is stable for a few seconds; the increased interface and residence time leads to complete carbon dioxide absorption even at a high flow rate of the carrier gas, while the latter would have to be slow without such a foam formation. The carrier gas (compressed air or nitrogen), freed from $CO_2$ on soda lime, is started only after decomposition of the organic material has terminated after 10 min. A carrier gas flow of 300-500 ml/min is conducted through for another 10 min with the formation of a foam zone of 2-4 cm height in the wash bottle. Subsequently, the wash bottle is disconnected and, after introducing an extended buret tip through the top opening and adding a few drops of phenolphthalein, the contents are

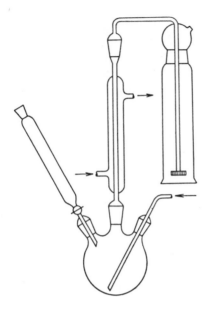

Fig. 5.5.3.2 Apparatus for the combined determination of the COD value and organic carbon content.

titrated with drop-wise use of 0.1 N HCl with vigorous shaking up to permanent decoloration. Near the end the contents of the frit are rinsed out once more under a slight vacuum. The same procedure serves for the determination of a blank value: 1 ml of 0.1 N HCl (blank value - sample titration) corresponds to 0.6 mg C.

Model solutions with 1 l of pure substances found in residential and in some industrial waste effluents led to COD values of $98 \pm 3\%$ and carbon values of $97 \pm 3\%$ of theory with the use of this simultaneous method.

Recently, greater attention is being given to the COD/org. C factor (see Table 5.3.4) in surface and waste waters (Sontheimer[294]; see also Schaffer and 6 co-workers[271]).

Naturally this factor is subject to considerable errors in the case of relatively clean water both because of the scatter of COD values as well as the fact that the organic C content appears as a small difference compared to a high concentration of total and inorganic carbon.

### 5.5.3.3    Determinations of the $NH_3$ and Amino Nitrogen Following the COD Determination

The knowledge that ammonia as well as amino, amido and nitrile nitrogen are finally present in the form of ammonium sulfate in the completely reacted $K_2Cr_2O_7$ oxidation solution suggests that the digestion solution should be divided into one aliquot for titration of the COD value with Fe(II) solution and another for a nitrogen determination (Leithe[189]).

For this purpose the digestion solution is rinsed into a 250 ml graduated flask, brought to the mark with distilled water and 200 ml are removed for back-titration of the chromate excess. The titration result is multiplied by the factor 1.25. Another 25 ml sample is pipetted from the graduated flask into a customary ammonia distillation apparatus with drop collector, treated with 50-60 ml sodium hydroxide solution and 50 ml water, and is distilled on a small open gas flame (with a few glass beads as a boiling aid) until all of the $NH_3$ has been driven over into a receiver consisting of a known quantity of excess 0.01 N $H_2SO_4$ weakly stained with methyl red. The main quantity of excess acid is back-titrated with 0.01 N NaOH from a 10 ml microburet, heated to boiling, boiled about 10 sec, cooled again and titrated to the yellow color to which the blank value was adjusted.

Naturally, the water used for dilution as well as the reagents have to be free from $NH_3$ or nitrogen compounds which can decompose into $NH_3$. The soda solution (300 g NaOH analytical grade $+$ 30 g $Na_2S_2O_3$ per liter) is steam-refined before use until no Nessler reaction is obtained in the distillate. The remaining $NH_3$ quantities must be determined by blank-value analyses. One ml 0.01 N $H_2SO_4$ corresponds to 140 $\mu$g N in the distillate.

With lower nitrogen contents a colorimetric $NH_3$ determination according to Nessler or by means of the indophenol reaction (see also Leithe and Petschl[199a]) is performed in the usual manner instead of an alkalimetric titration.

Model solutions with 17 of the pure substances expected to be present in residential and industrial waste effluents yielded COD values of 98 $\pm$ 3% and N-values of 98 $\pm$ 4% of theory according to the combined method.

### 5.5.4 Other Titration Methods with $K_2Cr_2O_7$ Determination of the COD Value with Sulfuric-Phosphoric Acid

According to Beuthe[26] 25 ml of the water sample with 20 ml of 0.01 N $K_2Cr_2O_7$ solution are boiled for 30 min in sulfuric acid diluted with the same volume of water and 30 ml of a mixture of equal volumes of concentrated $H_2SO_4$ and concentrated $H_3PO_4$. Subsequently, 0.3 g $Ag_2SO_4$ is added and boiling is continued for 90 min. After cooling, the product is diluted with water and titrated against ferroin with 0.1 N $Fe(NH_4)_2(SO_4)_2$. 0.23 mg/l COD should be subtracted per mg of Cl'.

*Rapid method according to Jeris.* Another rapid method for the COD determination with $K_2Cr_2O_7$ was developed by Jeris[152]. In the following it is described in the modified form of Wells[330]. Although the method is still somewhat more rapid than that described in 5.5.3, it is probably considerably inferior to the latter in sensitivity, reproducibility and completeness of the oxidation with a minimum blank value.

*Procedure:* A 125 ml flask with an inserted thermometer is charged with 0.3 g $HgSO_4$, 10 ml water sample and 25 ml acid dichromate solution and heated to $165 \pm 1°C$ on a preheated hot plate. The contents are allowed to cool to 80° and transferred into a 500 ml Erlenmeyer flask containing 250 ml water, cooled to 25-30° and titrated with 0.05 N Fe(II) solution against ferroin indicator. A blank value determination is made at the same time with 10 ml water. To adjust the titer of the Fe(II) solution 25 ml of aqueous 0.05 N $K_2Cr_2O_7$ and 20 ml concentrated $H_2SO_4$ acid are prepared in 250 ml water.

*Reagents:* 0.5 N acid dichromate solution: 5 g $K_2Cr_2O_7$ and 20 g $Ag_2SO_4$ in a mixture of 1 l concentrated sulfuric acid and 1 l concentrated phosphoric acid. 0.05 N Fe(II) solution: 20 g $Fe(NH_4)_2(SO_4)_2 \cdot 6 H_2O$ + 20 ml conc. $H_2SO_4$ brought to 1 l solution with water.

0.05 N $K_2Cr_2O_7$: 2.452 g to 1 l with water.

Foulds and Lunsford[91] find lower values according to Jeris than according to the APHA standard method.

### 5.5.5 *Photometric COD Determinations with* $K_2Cr_2O_7$

It seems indicated to determine the $K_2Cr_2O_7$ consumption after boiling of the water sample by photometry of the green Cr(III) ion formed or the unconsumed $Cr_2O_7^{1-}$. Both methods have been used and led to an automation of the COD determination. The minimal boiling time as described in the procedure of Leithe (5.5.3) offers particular advantages here.

The calibration curve for measurement of the formed green Cr(III) ion can be constructed in a simple manner with oxalic acid solutions of which 20.0 ml are reacted with 0.25 N or 0.05

N $K_2Cr_2O_7$ solution for 10 min according to the procedure of 5.5.3. The fully reacted solutions are rinsed into a 100 ml graduated flask and brought to the mark. The absorbance values obtained with it at 610 nm in the 5 cm cuvette of the MPQ II Zeiss spectrophotometer are listed in Table 5.5.5.

Automatic equipment for the photometric COD determination with $K_2Cr_2O_7$ for semi-continuous as well as continuous operation has been described.

**Table 5.5.5.—Standard photometric solutions**

| mg oxalic acid/l ($C_2H_2O_4 \cdot 2\ H_2O$) | COD/l | Absorbance | Factor |
|---|---|---|---|
| Determination with 0.25 N $K_2Cr_2O_7$ | | | |
| 6300 | 800 | 0.640 | 1250 |
| 4730 | 600 | 0.478 | 1250 |
| 3150 | 400 | 0.326 | 1210 |
| Determination with 0.05 N $K_2Cr_2O_7$ | | | |
| 1260 | 160 | 0.133 | 1200 |
| 946 | 120 | 0.102 | 1200 |
| 630 | 80 | 0.068 | 1180 |
| 315 | 40 | 0.040 | 1000 |
| 157 | 20 | 0.020 | 1000 |
| 78 | 10 | 0.012 | 1200 |

*Semi-continuous apparatus of the Johnson Company.* In the apparatus of Johnson Company (Nynäshamn, Sweden) (Fig. 5.5.5), a water sample of 8 ml is treated with mercury sulfate solution, adjusted $K_2Cr_2O_7$ solution and $Ag_2SO_4$-containing concentrated sulfuric acid by means of metering piston pumps and magnetic valves and heated to boiling for a preset period of time by means of an electrical heater. The green color of the solution which forms is continuously read with largely monochromatic light. After completion of the planned reaction period, the discharge valve opens, the system is flushed with water and newly filled. The zero drift corresponds to a COD value of less than 0.5 mg/l and the reproducibility of optical readings referred to full deflection for 250 mg/COD/l is reported to be $\pm$ 2 mg/l.

It may be assumed that the Silicometer of Bran and Lübbe Company might also be suitable for the present purpose after appropriate adaptation.

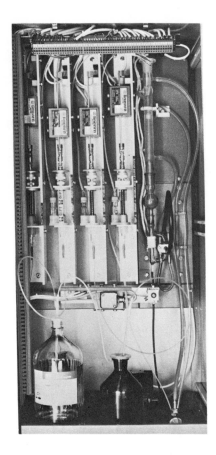

Fig. 5.5.5 — Semi-continuous apparatus (Johnson Company).

*Continuous method.* A fully continuous COD determination has the advantage that a test value is available at all times but requires a larger amount of reagents.

The experiments and results concerning its application were described by Hey, Green and Harkness[135]. The continuously metered sample mixture corresponding to 63.5 vol.% sulfuric acid in the presence of $HgSO_4$ and $Ag_2SO_4$ is heated to 160° in a long glass coil in a sand bath with a residence time of about 30 min. The photometric reading at 440 nm shows the residual $CrO_4$ ion. The reproducibility of a larger series of readings with waste water having a COD value of 90 amounted to ± 6 mg/l. Compared to the results of the APHA standard methods, the mean deviations can be calculated to be ± 8%.

The use of the Technicon autoanalyzer for photometric COD determinations has been described repeatedly. Zaleiko and

Molof[345] measure the residual $CrO_4$ excess, while Wagner[321] reads the formed green $Cr(III)$ ion (see also Adelmann[5a]).

### 5.5.6 *Automatic Potentiometric Titration of the COD Value*

In place of visual titration, the $K_2Cr_2O_7$ consumption can also be determined by potentiometric titration. Baughman, Butler and Sanders[21] use the Fisher automatic titralyzer for this purpose.

A water sample of 20 ml with 5.0 ml 0.03 N $K_2Cr_2O_7$ solution, 25 ml $Ag_2SO_4$-containing concentrated sulfuric acid and a few glass beads is heated for 2 h with refluxing. The solution is diluted to 100 ml and the unconsumed chromate is titrated potentiometrically (tungsten reference elecrode, platinum indicator electrode) with 0.025 N $Fe(II)$ solution in the Fisher automatic titralyzer. Before the analysis the electrodes must be rinsed with dilute HCl. Coefficient of variation $\pm$ 5%.

### 5.6 COD Determination with Potassium Persulfate by Gas Volumetry

As described in 5.3.3.2, organic substances can be oxidized completely into $CO_2$ and $H_2O$ with potassium persulfate ($K_2S_2O_8$) in the presence of a silver catalyst. As demonstrated by Leithe[188] this reaction can also be utilized for a COD determination by determining the consumption of oxidant instead of the formed $CO_2$. However, this oxidant may not be used in the otherwise customary large excess; nor is this necessary for most organic water pollutants (carbohydrates, amino acids, carboxylic acids) and usually approximately double the theoretical quantity of $K_2S_2O_8$ is sufficient.

While $K_7Cr_2O_2$ is sufficiently stable when used as an oxidant for the COD determination that the remaining excess can be back-titrated, this is not the case for $K_2S_2O_8$ since a considerable self-decomposition with oxygen evolution takes place simultaneous with the oxidation of organic material. However, it proved to be possible to determine the oxygen formed by self-decomposition together with the unconsumed quantity of $K_2S_2O_8$ by gas volumetry: thus, subsequent to oxidation of the water sample the $K_2S_2O_8$ remaining in the excess is decomposed by further boiling with oxygen formation, the entire oxygen evolved is driven into an azotometer with $CO_2$ and read in the latter. If a blank decomposition with clean water is carried out with the same quantity of $K_2S_2O_8$, the difference of the $O_2$ volumes can serve as a criterion for the oxygen consumption to oxidize the organic water pollutants and thus to determine the COD value.

For nitrogen-containing materials, a basic and significant difference of the reaction of $K_2S_2O_8$ and $K_2Cr_2O_7$ has been observed. While the nitrogen of $NH_3$ and of amino compounds is not oxidized with $K_2Cr_2O_7$ but the ammonium sulfate is finally obtained, it is almost completely converted into nitrate with $K_2S_2O_8$ in the presence of silver. The quantitative conversion of ammonia and amino nitrogen into nitrate with a subsequent oxidimetric nitrate determination has been developed into a new method of nitrogen determination in such materials (Leithe[194]).

The nitrate formation is primarily attributable to the action of silver, since in its absence, ammonium sulfate is essentially formed also with $K_2S_2O_8$ according to the reports of Winkler (see 5.10.1).

For a water analysis, the nitrate formation is important insofar as the COD value of nitrogen-containing compounds obtained with $K_2S_2O_8$ is higher by the amount of oxygen required to convert the nitrogen into nitrate than the COD value found with $K_2Cr_2O_7$. It is therefore suitably identified by $COD_{NO_3}$. Thus, it corresponds to the oxygen demand to be expected in waste water technology with complete nitrification of the nitrogen compounds during intensive treatment in the activated sludge process.

The present $COD_{NO_3}$ method is more rapid than the 2-h method with $K_2Cr_2O_7$, but more tedious than the 10 min method described in 5.5.3 and also requires more manipulations. Automation of the oxygen determination (see, for example, 5.7) might perhaps eliminate these problems. Practical results with residential or industrial waste water have not been published thus far.

*Procedure* (see Fig. 5.6 for apparatus). A Dewar jar filled with dry ice serves for the recovery of air-free $CO_2$. To increase the gas pressure, the jar can be immersed in warm water or the vacuum in the jar can be broken by cutting off the glass tip. The charge is sufficient for about one day. It is also possible to use the Kipp apparatus prepared according to Pregl for an elemental microanalysis according to Dumas.

The decomposition flask is a 250 ml ground-joint round-bottom flask with dropping funnel and Liebig condenser. A few glass beads serve as a boiling aid.

The azotometer of 50-100 ml volume is filled with 50% KOH. For rapid and precise stabilization the buret is surrounded by a

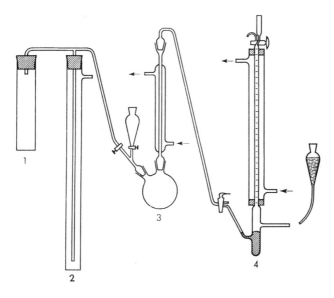

Fig. 5.6—Apparatus for COD determination with $K_2S_2O_8$; azotometer method. (1) Dewar jar for $CO_2$; (2) protective bath; (3) decomposition flask; (4) azotometer.

transparent cooling jacket and cooling water is obtained from a circulating thermostat.

The apparatus is first purged with air-free $CO_2$ until micro-bubbles appear in the azotometer. The $CO_2$ stream is then disconnected, the decomposition flask is charged through the dropping funnel, avoiding entry of air (overpressure with a small attached rubber blower) with 100 ml of the water sample, 25.0 ml 0.3 N $K_2S_2O_8$ solution (40.6 g $K_2S_2O_8$/l; Merck, nitrogen-free grade), 5 ml 10% $AgNO_3$ solution and 2 ml 2 N $H_2SO_4$. A gentle $CO_2$ stream is started, the solution is slowly heated to boiling with a small flame (about 5 min up to the start of boiling), boiling is continued for about 5-10 min in a stronger $CO_2$ stream until the gas bubbles have become microbubbles again; the temperature and air pressure are then read and the read volume is converted to standard conditions. The $O_2$ volume, obtained by exactly the same method from 100 ml clean deaerated water, 25 ml 0.3 N $K_2S_2O_8$ solution and the above reagents, serves as the blank value.

In order to obtain the COD value from the $O_2$ volume difference obtained (blank value minus analysis), a calibration curve is constructed with oxalic acid solutions with COD values of 50,

150, 300 and 400 mg/l (1 mg COD/l corresponds to 7.87 mg oxalic acid $C_2H_2O_4 \cdot 2\ H_2O$/l).

Interferences: Chlorides already interfere in low concentration (starting at 10 mg Cl⁻/l) and can be removed as follows: a 100 ml graduated flask provided with a mark at 110 ml is filled with 100 ml water sample which is treated with 2 ml 2 N $H_2SO_4$ and 5 ml 10% $AgNO_3$ solution. The sample is brought to the 110 ml mark and is allowed to stand a few minutes until the precipitate forms, is filtered and 100 ml of the filtrate is pipetted for analysis. The result must be multiplied by 1.10.

*Results:* In model waste water samples from 13 different organic materials such as those expected to be present in residential and industrial waste effluents, the COD values obtained corresponded on the average to 98 ± 3% of theory (including nitrification). The standard deviations in parallel determinations amounted to ± 0.2 ml of the measured $O_2$ volumes. Only about 50% of the urea is determined with $K_2S_2O_8$ with nitrate formation.

## 5.7    Automatic Dry Methods for COD Determination

*Automatic COD determination by oxygen measurement in the combustion atmosphere.* The Ionics Company of Watertown, Massachusetts is marketing a total oxygen demand analyzer, Model 225, developed by Dow Chemical Company. The water sample (10 µl, see 5.3.2.1-2 is evaporated in a nitrogen stream and combusted with metered quantities of oxygen at 900° on a platinum screen catalyst. The reduction of the original oxygen concentration is measured with a galvanic platinum-lead cell in the form of a depolarization current. The duration of an analysis is 3 min and the measuring range extends up to 1000 mg/l; the manufacturer reports a reproducibility of ± 2% of the measuring range.

A basic difference compared to the results of methods with $K_2Cr_2O_7$ consists of the fact that the nitrogen present in the water sample is completely converted into NO according to claims of the manufacturer. Accordingly, it may be expected that the values obtained with this apparatus, which might be logically identified as $COD_{NO}$- values (see Leithe[108]) would be higher than the COD values with $K_2Cr_2O_7$.

Clifford[52] has reported unfavorable results with this method.

*Automatic determination of the COD value by conversion of* $CO_2$ *into CO*. Stenger and Van Hall[297] have described a 2-min method in which 20 $\mu$l of water sample is evaporated in a $CO_2$ stream and conducted over a platinum screen heated to 875°. In this process the organic materials reduce $CO_2$ into CO which is determined on a non-dispersive CO-sensitized IR spectrometer.

A similar apparatus, the "Aquarator," is manufactured by Precision Scientific Co. (Chicago) and marketed in Germany by the Classen Company (Hamburg). It is designed for the determination of COD values from 10 to 300; the reproducibility of measurements is reported to be ± 3%.

In these procedures the nitrogen bound in the water samples is separated in elemental form, so that plus-values can be expected with this method compared to oxidation with $K_2Cr_2O_7$. Nitrate in the sample as well as nitro-compounds release their oxygen content and sulfate is reduced into sulfite. According to Stenger and Van Hall, a factor of $COD_{N_2}:COD$ (with $K_2Cr_2O_7$) = 1.10 ± 0.07 may be expected for waste waters. With regard to the nomenclature, see Leithe[198].

## 5.8 Evaluation of Water Samples on the Basis of the COD Value

Although the COD value of a water sample can be considered a criterion of the total organic pollutants, it does not allow their quantitative calculation any more than the organic carbon content if the elemental composition is not known. The conversion factor changes as a function of the oxygen and nitrogen content of the respective materials as demonstrated by Table 5.8a. In domestic waste water and effluents of similar composition, a guideline value of about 1.2 mg COD per mg of substance may be accepted.

Table 5.8a—COD-values, per g of substance

| | | | | | |
|---|---|---|---|---|---|
| Oxalic acid | 0.18 | Peptone | 1.20 | Butyric acid | 1.82 |
| Hydrocyanic acid | 0.59 | Albumin | 1.32 | Ethanol | 2.09 |
| Glycocoll | 0.64 | Casein | 1.39 | Phenol | 2.38 |
| Glucose | 1.07 | Sulfite pulp waste | 1.32 | Butanol | 2.60 |
| Acetic Acid | 1.07 | | | Na dodecyl- | |
| Saccharose | 1.12 | | | benzenesulfonate | 2.35 |

The literature frequently contains discussions on whether the rapid and simple COD determination might serve as a substitute for the considerably more prolonged and tedious $BOD_5$ determination. Within certain limits this is possible in many cases, particularly when only pollutants with good biodegradability are present. According to earlier findings of Theriault with fresh residential waste water, approximately 70% degradation occurs after 5 days in the BOD determination $(COD = 1.4\ BOD_5)$[1]. The balance has not been attacked or has been converted into new biomass. A complete degradation can be expected only after 20 days $(COD = BOD_{20})$. However, if pollutants with sparing biodegradability are present, this factor increases considerably. Such materials may be of industrial origin (sulfite pulp effluent, sparingly biodegradable protein degradation products from leather factories, etc.), or involve residues in the effluents from biological treatment plants in which the more easily biodegradable constituents have already been removed.

An evaluation of these sparingly biodegradable residual materials, the presence of which becomes evident from an increase of the $COD/BOD_5$ factor of 1.4 will be possible only on the basis of additional analyses for their chemical nature and expected effect. Undoubtedly, in the case of higher concentrations and prolonged retention times they will also have an influence on the oxygen balance of the sewer.

In recent years, the ratio mg $COD$/mg org. C has also become of some interest (Sontheimer[294], as well as Schaffer et al.[271]). In the case of pollutants which are mainly expected in residential effluents, it amounts to about 2.7 according to Table 5.3.4, while with materials of higher carbon and hydrogen content, such as those expected in the pollutants of chemical synthesis, it may attain 4.

The COD values of clean water which are considered for drinking water are of special interest. It was already indicated in 5.4.1.5 that with the use of $K_2Cr_2O_7$ the organic pollutants are determined completely, while $KMnO_4$ used for the customary determination of the $KMnO_4$ consumption attacks only about 25% of the oxidizable components. The COD value of such water can thus serve as a more reliable measure of its oxidability which, for

---

[1] There is an inherent problem in stating expected $BOD_5/COD$ relationships for various kinds of watse water. For example, "residential wastes" in Europe and the U.S. vary as a function of life style alone. They are a function of their contents. Household products which are available in the U.S. may not be available in Europe and vice versa. Even without reference to the location, this factor $(BOD_5/COD)$ should be expected to change as technology changes and new products are made available to the housewife.

example, may become manifest in an increased requirement of ozone or chlorine for drinking water disinfection.

Relationships between the COD, $BOD_5$ and $KMnO_4$ values and the saprobic stages could also be demonstrated in a few rivers of Upper Austria (Leithe[199]), even though differences would be expected on the basis of the type of pollutants (Table 5.8b).

Table 5.8b.—Relationship between saprobic stage, COD, $BOD_5$ and mg $KMnO_4$-values in rivers of Upper Austria

| Saprobic stage | COD (mg/l) | $BOD_5$ (mg/l) | mg $KMnO_4$/l |
|---|---|---|---|
| I | 2 | 2 | 6 |
| I/II | 4 | 4 | 6 |
| II | 8—9 | 3—4 | 11—15 |
| II/III | 11—18 | 4—7 | 26—35 |
| III | 20—65 | 20 | 30—150 |
| III/IV | 80—200 | 40—120 | 150—390 |

## 5.9    Determination of the Chlorine Value According to Froboese[95]

Another oxidant which was frequently used in the past to characterize the oxidizability of a water sample is sodium hypochlorite. DEV, First Edition, lists the procedure discussed below.

The water sample of 100 ml is brought to boiling with 20 ml N/50 NaOCl solution, which should require about 5-5½ min. Boiling is continued for exactly 10 min, followed by cooling, addition of 1 g KI and 1 ml $H_3PO_4$ (25%) and the liberated iodine is titrated 10 min later with N/50 $Na_2S_2O_3$ solution as customary. One ml of the consumption difference (20 ml $Na_2S_2O_3$) corresponds to 0.709 mg chlorine.

The N/50 NaOCl solution is prepared by diluting commercial sodium hypochlorite solution. Its alkalinity should correspond to 0.1 N NaOH. For its determination, 20 ml neutral $H_2O_2$ solution are added to 25 ml of NaOCl solution diluted to N/50, and after 10 min the solution is titrated with 0.1 N HCl against methyl orange.

Sodium hypochlorite is a very selective oxidant which reacts preferentially to ammonia, urea, amino acids and the like, but less to carbohydrates. In this regard, a determination of the chlorine number completes that of the permanganate consumption and occasionally gives information on the origin of a waste effluent. Reaction in this direction becomes more selective if it is carried out for only 1 min and if for more highly polluted waters, 0.1 N solution is used instead of N/50 NaOCl. This weakens the action on carbohydrates, while that on nitrogen compounds remains unchanged (Leithe[197]).

## 5.10   Organic Nitrogen Determination

Two fairly old methods are available for the determination of "organic nitrogen" in order to analyze for water-soluble intermediates of protein degradation and thus demonstrate the presence of insufficiently mineralized organic wastes. However, both methods are subject to considerable possibility of error and are therefore rarely used at this time. The alternate consists of the Kjeldahl total nitrogen determination (possibly in the absence of nitrate without addition of reducing agents) or according to 5.5.3.3 and subtraction of the ammonia nitrogen which has been distilled over with weak alkalinity (addition of MgO or dilute soda).

### 5.10.1   Determination of "Protein Ammonia" with Potassium Persulfate

This method dates back to Winkler[339]. According to DEV (First Edition), 100 ml water sample is slightly acidified with $H_2SO_4$, treated with 1 g $K_2S_2O_8$ and maintained near boiling on a water bath for 15 min. After cooling, 2 ml 50% Rochelle salt solution and 2 ml Nessler reagent are added. A second sample is treated in the same manner but without persulfate and sulfuric acid. It shows the quantity of free ammonia nitrogen present. An ammonium chloride reference solution (3.0 g $NH_4Cl$ per 1 with $NH_3$-free water; 1 ml = 1 mg $NH_4$) is added dropwise from a buret until the color of the two samples is equivalent. The $NH_4Cl$ consumption yields the quantity of "protein ammonia."

### 5.10.2   Determination of "Albuminoid Ammonia"

An alkaline permanganate solution decomposes protein and albuminoid solutions with ammonia formation. However, the reaction is incomplete, *i.e.*, only about 50% according to older data.

According to Ohlmüller-Spitta, the "free ammonia" is first determined by distilling 200 ml from 100-250 ml water sample which has been diluted with 500 ml $NH_3$-free water and treated with 1 g MgO or 0.5 g $Na_2CO_3$. Subsequently, 50 ml alkaline $KMnO_4$ solution are added and an additional 100 ml are distilled away. In the next 50 ml of distillate, the test is performed with Nessler reagent; if the reaction is still positive, another 50 ml are distilled away. The $NH_3$ concentration of the individual distillates is determined with Nessler reagent.

According to the APHA standard methods, 1000 ml of water sample are treated with 10 ml phosphate buffer of pH $= 7.4$ (14.3 g $KH_2PO_4$ + 68.8 g $K_2HPO_4$ per liter) and distillation is performed until the distillate shows a negative Nessler reaction. Subsequently 50 ml of alkaline $KMnO_4$ solution are added, 200-250 ml distillate are collected and the $NH_3$-content in the latter is determined with Nessler reagent.

Alkaline $KMnO_4$ solution: 1200 ml distilled water are heated to boiling in a 2½ l porcelain dish. Then 16 g $KMnO_4$ are dissolved in it, 800 ml 50% KOH solution are added, the vessel is filled with distilled water and the solution is evaporated to 2 l. The blank value from 50 ml diluted with 50 ml water and distilled must be taken into account.

## 5.11 Determination of the Biochemical Oxygen Demand ($BOD_5$)

The biochemical oxygen demand (BOD) refers to the quantity of elemental oxygen in mg which is bound by 1 h of water with the participation of microorganisms in the biochemical degradation and transformation of the organic pollutants of a water sample. In the practice of water conservation, the value determined under standard conditions serves as a criterion of the total organic pollutants accessible to biochemical degradation in a body of water or a waste water sample.

A precursor of the $BOD_5$-value was the *oxygen consumption* according to Spitta, which has been used primarily for surface waters. Two identical samples of the water to be tested are filled each into a suitable stoppered glass bottle without an air bubble in the manner prescribed for the oxygen determination. The dissolved oxygen of one sample is determined immediately; the second sample is stored in the absence of light at 20° and the residual oxygen is determined after 48 h. The difference in mg $O_2/l$ yields the "oxygen consumption."

With the extension of the method to more highly polluted water, particularly waste water, the quantity of oxygen directly dissolved in water is not sufficient and requires the addition of a known quantity. Furthermore, because of the fairly long incubation time of bacterial processes, which is frequently necessary, the standing time needs to be prolonged (as a rule to 5 days). In this form the result obtained in mg $O_2/l$ is known as $BOD_5$.

As a rule, therefore, the $BOD_5$ determination is made by the addition of a known and sufficient quantity of elemental oxygen in the form of $O_2$-saturated dilution water or as a metered volume of air or $O_2$, and at the end of the contact time, a residual amount of at least 2 mg oxygen/l is necessary.

However, a $BOD_5$ determination can be carried out not only on the basis of the oxygen consumption but also on that of the decrease of organic matter as a result of 5 days of bacterial activity; Leithe[195] has demonstrated that this value is obtained from the difference of the COD values before and after 5 days.

Every type of $BOD_5$ determination requires the presence of bacterial flora capable of degrading the organic matter present in the sample. Depending on the objective of the test, inoculation is performed with water of the respective sewer, with fresh or stale residential waste water or with the waste effluent of an operational biological treatment plant. In the case of industrial waste effluents with exogeneous pollutants, very different results may be expected depending on whether the available bacterial strains respond to the respective components or not.

If the holding time of 5 days has not been observed for technical reasons, the $BOD_n$-value obtained after n days can be converted into the $BOD_5$ value by the factor listed in Table 5.11 (according to Imhoff[144a]).

Table 5.11—Conversion factors for $BOD_n$ into $BOD_5$

| n days | 2 | 3 | 4 | 5 | 6 | 7 | 8 |
|---|---|---|---|---|---|---|---|
| $BOD_n/BOD_5$ | 0.54 | 0.73 | 0.88 | 1.00 | 1.10 | 1.17 | 1.23 |

### 5.11.1  Methods for Determination of the Oxygen Consumption

The following methods will be discussed on the basis of the method of supplying the measured quantity of oxygen, the decrease of which is to be determined:

1. Dilution methods
   a) With air-saturated dilution water
   b) With pure oxygen saturation

2. Manometric methods. The water sample in a closed container is treated with a metered volume of air and the pressure

drop produced by the decrease of oxygen is determined mano-metrically. The carbon dioxide formed during treatment is bound on solid caustic alkalis. If necessary a metered quantity of pure makeup oxygen is added. Proposed equipment consists of the customary Warburg apparatus, an enlarged Sierp-Fränse-meyer apparatus as well as recently marketed glasses with magnetic stirrers and membrane manometers set up in batteries.

3. Coulometric methods. The oxygen consumed by bacterial processes is supplemented by electrolysis. The quantity of current required for this purpose serves as a criterion for the formed oxygen.

Montgomery[225] has surveyed methods of $BOD_5$ determination.

*Sample preparation.* Depending on the intended objective, the water sample is filtered or homogenized.

In order to guarantee adequate bacterial action the pH of highly alkaline or acid samples must be adjusted to 7-8.

In the presence of materials which consume elemental oxygen even without the presence of bacteria, for example, Fe(II) com-pounds, $H_2S$ or $SO_2$, the aerated sample is allowed to stand for 2 h if the determination is to be limited to biochemical effects.

Free chlorine is bound by the addition of sodium thiosulfate in the necessary quantity.

In the case of sterile samples or samples with a low bacterial count, the sample is inoculated with 0.3 ml sedimented residential waste water, 2 ml biologically treated waste effluent or 5-10 ml river water, preferably from the appropriate sewer, per liter de-pending on the purpose of the test.

### 5.11.1.1 $BOD_5$ Determination by the Dilution Method

Apparatus: Oxygen bottles of about 100 or 250 ml volume, pre-cisely measured with identical labeling of the preferably somewhat beveled ground stopper. (EDITOR'S NOTE: 300 ml volume is common in U.S.)

Dilution water: clean "depleted" water with a $BOD_5$ value of less than 1 saturated with atmospheric oxygen. Per liter, it is treated wtih the following quantities of buffer and nutrients:

8.5 mg $KH_2PO_4$, 21.75 mg $K_2HPO_4$, 33.4 mg $Na_2HPO_4 \cdot 2 \ H_2O$, 1.7 mg $NH_4Cl$, 22.5 mg $MgSO_4 \cdot 7 \ H_2O$, 27.5 mg $CaCl_2$, 0.25 mg $FeCl_3$.

Reagents for iodometric oxygen determination:

Mn(II) sulfate solution: 800 g $MnSO_4 \cdot 4 \ H_2O$ and 1 l distilled water.

Precipitation reagent: 360 g NaOH, 200 g KI and 5 g $NaN_3$ (caution—sodium azide) are dissolved with distilled water to form 1 l.

$H_2SO_4$, d = 1.70.

0.01 N $Na_2S_2O_3$ solution which must be standardized daily with 0.01 $KMnO_4$ solution or with $K_2Cr_2O_7$. (A preservative such as chloroform can be added.)

Zinc iodine starch solution: Four g of starch are suspended in a small amount of water and poured into a boiling solution of 20 g $ZnCl_2$ in 100 ml water[2]. The mixture is boiled until it becomes clear, is diluted, 2 g $ZnI_2$ are added, followed by dilution to 1 l and, if necessary, by filtration.

Dilution of water sample: sufficient dilution water must be added so that a residual concentration of at least 2 mg $O_2$/l can be found at the end of the holding time.

In the case of residential waste effluents the necessary dilution can be estimated from the $KMnO_4$ consumption according to Table 5.11.1.1.

Table 5.11.1.1

| Potassium permanganate consumption (mg $KMnO_4$/l) | Expected $BOD_5$ (mg/l) | Necessary water sample in ml to be diluted to 1000 ml |
|---|---|---|
| to   15 | to   10 | 250 and 150 |
| 15— 40 | 10— 30 | 100 and  75 |
| 40— 60 | 20— 50 | 50 and  40 |
| 60—120 | 40—100 | 30 and  20 |
| 120—240 | 80—200 | 15 and  10 |
| 240—360 | 160—300 | 10 |

In effluents from biological treatment plants in which the easily biodegradable materials have been removed preferentially a considerably lower $BOD_5$ value may be expected. The same is true for surface water in which the more easily biodegradable pollutants are no longer available after longer retention times. For industrial waste effluents the above relation is not valid at all.

Mixing of the water sample with dilution water may be carried out in graduated flasks of 500-1000 ml volume; 2-3 oxygen bottles are filled with the mixture without air bubbles. However, the measured water sample can also be directly charged into the

---

2. Used with potato starch, salicylic acid is recommended preservative.

oxygen bottles and brought to the mark with dilution water without a loss of sample water. In one of the 2-3 filled bottles the $O_2$ content is determined immediately, and in the other or others after 5 days of standing (in darkness a 20°). At the same time one bottle is filled identically with inoculated dilution water alone and tested after 5 days.

Oxygen determination: Suitable pipettes serve to introduce 0.5 ml (or 1 ml for 250 ml bottles) of $MnSO_4$ solution and an equal volume of precipitation reagent into the opened bottle. The bottles are immediately closed and shaken well by inversion. After settling of the precipitate the bottle is opened and 2 ml $H_2SO_4$ are added from a pipette. The bottles are closed again, shaken to form a solution, allowed to stand for a few minutes and titrated with 0.01 N $Na_2S_2O_3$ solution until colorless—near the end of the titration with the addition of about 1 ml zinc iodine-starch solution.

Editor Note: Common U.S. method is to shake by inversion the second time after the precipitate from the first mix has settled. Acid is added only after the precipitate (or flocculant) has settled the second time.

Calculation:

$$\text{Oxygen content (mg } O_2/l) = \frac{\text{ml bottle vol.} - \text{ml reagents}}{\text{ml } 0.01 \text{ N } Na_2S_2O_3 \cdot 80}$$

$$\text{Consumption } Z = \text{mg } O_2/l \text{ immediately} - \text{mg } O_2/l \text{ after 5 days}$$

$$BOD_5 = \frac{d}{e} \cdot (Z_p - Z_v) + Z_v$$

$d$ = volume of graduated flask or bottle volume — ml reagents
$e$ = applied volume of undiluted sample
$Z_p$ = consumption of dilution
$Z_v$ = consumption of dilution water.

With dilutions of 1 + more than 99 the formula simplifies into:

$$BOD_5 = \frac{d}{e} (Z_p - Z_v)$$

*Electrochemical oxygen determination.* In place of the titrimetric oxygen determination according to Winkler, which is tedious (see, for example, Montgomery and Cockburn[226]), an automatic determination by a galvanometric or polarometric method has recently become common. As a rule, electrode systems are used which are coated with a selective membrane allowing the permeation of elemental oxygen but not of most foreign ions. (There may be interferences.) The electrodes are either a gold cathode and silver anode, the cathode being polarized by an applied voltage of 0.8 V, or a combination of noble metal cathode/ lead anode without the need for an applied voltage is used. In both cases, the oxygen present causes a depolarization and therefore a current flow which is proportional to the present concentration of elemental oxygen and is indicated on a scale or recorder in mg $O_2$/l. For the measurement the electrodes are simply immersed in a flowing water sample and a reading can be taken after a few seconds. (Flow can be achieved in the laboratory with a magnetic-mixer, but flow past the electron is necessary.) As a rule, an automatic temperature correction is provided. The accuracy amounts to 1-3% of the measured value. Equipment of this type is marketed by numerous firms, for example, WTW Weilheim, Beckman-Instrument, Yellowsprings (polarographic) and Precision Scientific (galvanic), U.S.

The method is also described by Von der Emde and Kayser[320], Fastabend and Handloser[82], Schmid and Mancy[275].

### 5.11.1.2  Method with Saturation Via Pure Oxygen

Sierp[286] suggested that the water sample for a $BOD_5$ determination be saturated with pure oxygen instead of air, so that about 5 times the quantity of oxygen can be introduced and the addition of dilution water becomes unnecessary, especially in the case of surface water. A separatory funnel is filled with the water sample, the air above it is displaced by pure oxygen and the funnel is shaken for about 1 min. Two $BOD_5$-analysis bottles are then filled for an immediate determination of the oxygen and its residual amount after 5 days. The quantity of reagents must be doubled for the oxygen determination.

Viehl[318] as well as Günther[118] report on good agreement compared to air saturation.

### 5.11.1.3  Warburg Method

Warburg apparatus: Thirteen ellipsoidal reaction vessels of 120-140 ml capacity each with a center cylinder (as well as suitable manometers).

Procedure: Forty ml waste water sample are pipetted into the vessels and 1 ml KOH solution (15 g KOH in 100 ml water) into the center cylinders and a water bath is adjusted to 20°. After temperature stabilization the manometers are adjusted to 15 cm, the cocks are closed and the shaker is started. Depending on the consumption rate, a reading is taken every 30 min to approximately every 24 h while the shaker is stopped. With higher consumptions the $O_2$ loss must be made up. The $BOD_5$ value is the sum of differences of the individual manometer readings and the reading of the corresponding control manometers multiplied by the corresponding vessel constants.

$$\text{Vessel constant } K = \frac{a}{b}\left[\frac{c \cdot d}{e \cdot f} + \frac{g \cdot x}{f}\right]$$

a = specific gravity of oxygen in mg/l (1429)
b = volume of waste water sample including dilution water
c = 273
d = total volume of vessel and manometer in ml minus b
e = 273 + experimental temperature in °C
f = standard atmospheric pressure in mb (1013)
g = sum of b and the volume of KOH solution
x = Bunsen adsorption coefficient (at 15°C 0.03415,
                                     at 20°C 0.03103,
                                     at 25°C 0.02831).

See Schuller[279] on a comparison of the results with those of the dilution method.

### 5.11.1.4 Sierp-Fränsemeyer Method
### (See Fig. 5.11.1.4 for apparatus)

*Procedure:* The reaction flask is charged with 200-400 ml waste water, diluted if necessary, through the tube inserted into the flask and the tube is closed. The lower mark of the eudiometer tube is adjusted with the barrier liquid (saturated common salt solution containing 10 g NaOH/100 ml). With the use of undiluted oxygen the empty apparatus is first purged with oxygen and only then is the water sample introduced. After adjusting the level and reading the volume, the shaker is started. A reading is taken approximately every 4-5 h after the formed $CO_2$ has been absorbed by repeated raising and dropping of the leveling bottle. In waste waters with a high consumption the amount of oxygen consumed needs to be made up periodically with pure oxygen.

The gas volume readings are converted to standard conditions as usual.

$$BOD\ (mg/l) = \frac{y \cdot 1429}{Z}$$

y = sum of read oxygen consumptions under standard conditions
Z = applied sample volume in ml.

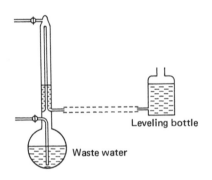

Fig. 5.11.1.4. — Apparatus according to Sierp and Fränsemeyer.

Leveling bottle

Waste water

### 5.11.1.5 Automatic BOD$_5$ Apparatus

*Barometric vibration BOD apparatus (Vedewa apparatus).* The equipment described by Burchardt[44] is a refinement of the methods of Sierp, Warburg as well as of the Stuttgart apparatus of Pöpel, Hunken and Steinecke[252]. The oxygen consumption is measured on a built-in aneroid barometer as an air pressure drop. The sample is aerated by a 50 Hz ac vibratory table on which the 6 units are mounted. With consideration of the volume of air space and of the charged sample, which will be selected larger the lower the expected BOD value, the scale deflection of the barometric reading furnishes the BOD result. The volumes utilized (162-575 ml) are those permitting the simplest conversion of the scale reading into BOD values. Temperature variations are compensated by a compensating device. Consideration must be given to the dissolved oxygen in the case of low BOD values and to the partial pressure of the formed $CO_2$ at higher values.

Standard deviations of $\pm$ 8.8% were found for BOD$_5$ values of 10-100 and of $\pm$ 6.9% for values of 100-800 (see also Sprenger[296]). Considerably higher deviations were obtained compared to the BOD dilution method as a consequence of the concentration dependence of the bacterial reactions.

A cyclic device for a BOD$_5$ determination in which the air is circulated by a membrane pump, the formed $CO_2$ is absorbed in a wash bottle and the oxygen decrease is measured by gas volumetry has been described by Pöpel, Hunken and Steinecke[252]. It is intended particularly for samples with a high $BOD_5$ value and high sludge content.

An instrument based on coulometry is the Sapromat of Voith Company, Heidenheim. Experiences with this apparatus have been described by Liebmann and Offhaus[203]. It consists of a battery of stirred vessels with magnetic stirrers. Each vessel holds a volume of 333 ml water sample. A volume of air maintained at constant pressure is above the sample: a $CO_2$ absorber removes the carbon dioxide formed. If the water sample consumes oxygen, a vacuum forms which produces electrical control pulses via a pressure indicator, which is independent of barometric-atmospheric pressure fluctuations and these pulses start an electrolytic oxygen evolution until the original pressure has been restored. The amount of current utilized for this process is indicated directly as mg $O_2$/l. The apparatus is built into a thermostat of 20°. After a preselected measuring time, the unit shuts off automatically. The process of oxygen consumption for 5 days is recorded in 120 measuring values. The scatter within parallel determinations amounts to $\pm$ 5%. In a comparison with the results of the $BOD_5$ determination by the customary dilution method, greater differences again occur because of the difference in concentration ratios, the values of the Sapromat often being up to 20% higher; the value of the $BOD_5$ dilution determination is frequently reached already after 48-72 h instead of after 120 h in the Sapromat.

Clark[49], as well as Young, Garner and Clark[343], Bridie[38] and Fuhs[49] have reported about other automatic coulometric $BOD_5$ analyzers.

Mullis and Schroeder[234] as well as Hiser and Busch[137] describe experiments intended to accelerate the BOD determination by stronger inoculation with bacterial cultures.

### 5.11.2  Determination of the $BOD_5$ Value by Difference Measurements of the Chemical Oxygen Demand (Leithe[195])

The $BOD_5$ value can be obtained not only by measuring the quantity of oxygen consumed in the process but also by determining the organic matter consumed by bacteria. The total quantity of original organic matter is characterized by the initial COD value and the balance of organic matter 5 days of bacterial activity in the presence of air is denoted by the COD value after 5 days. The difference yields the quantity of oxygen consumed during this period for oxidative degradation and is equal to the $BOD_5$ of the water sample.

The added oxygen or air does not need to be measured. It is sufficient for the water sample in a suitably large bottle with a sufficiently large air space to stand in darkness for 5 days at 20° with occasional shaking. More highly polluted water does not need to be diluted. The water sample is inoculated in the customary manner. Before sampling for the second analysis the water sample must be well shaken since it must contain the proportional amount of sludge newly formed in the elapsed time.

The new process was tested with $BOD_5$ determinations made by the dilution method. In the case of river and waste water samples which required little or no dilution for the dilution method the agreement was numerically good; however, if the new method was applied with undiluted waste water and the standard method in higher dilutions, somewhat greater differences resulted such as those which also occur with other methods using analogous dilution ratios (see 5.11.1.5 above).

The proposed $BOD_5$ analysis by measuring the COD difference requires no metering and measurement of the oxygen supplied to the water sample at the start and remaining after 5 days. The biochemical oxidation can take place in the undiluted state under natural conditions, and the manipulations and imprecisions of the customary dilution method are avoided. As an advantage compared to the above-described $BOD_5$ methods based on manometric or coulometric determinations, the low equipment costs should be pointed out.

With regard to relationships between the trend of the BOD values after 1-5 days, the simultaneous decrease of the COD value and the variation in bacterial count, see Kolaczkowski, Mejboum and Spandowska[168].

Difference measurements of the COD value as well as of organic carbon were utilized by Janicke[146] for an indirect determination of the biodegradability of nonionic detergents.

### 5.11.3 *Evaluation of Water Samples on the Basis of the $BOD_5$ Value*

In water and waste water chemistry the $BOD_5$ value is used as a criterion of the biodegradable pollutants present in a water and the resulting load on the oxygen balance of a body of water. In residential waste water, according to Theriault, it amounts to about 70% of the corresponding COD value which indicates the volume of oxygen required for complete oxidation of all organic pollutants as described in 5.8. That section gives a full discussion of the $BOD_5/COD$ factor which may assume considerably lower values than 0.70 in biologically pretreated as well as in many industrial waste effluents.

The relations between the $BOD_5$ value and the oxygen balance of a body of water have been formulated mathematically (equations of Streeter and Phelps) and discussed on the basis of empirical values by numerous authors (for example, Imhoff, Fair, Kehr, Pöpel, Böhnke).

Relations between the saprobic stages and $BOD_5$ values are listed in Tables 1.2.4 and 5.8b.

The table in the appendix gives the normal requirements for waste water treatment processes with regard to the desirable end $BOD_5$ values.

## 5.12    Physical Methods for Total Water Pollutant Indication

Various physical methods have been proposed for an overall characterization of the presence of organic water pollutants and an estimate of their concentration. Since these always refer only to certain properties of these materials, they do not have general and all-inclusive validity; however, provided they can be measured rapidly, simply and reliably, they can serve for a rapid indication of sudden organic pollutant loads in a body of water.

*UV absorption.* See Ogura and Hanya[242] concerning approximate measurements of water pollutants in the UV region at 220 nm. The absorption increases in parallel with an increasing COD value (see also Armstrong and Boalch[12]). In particular, humic acids and lignins are determined at 280 nm (Mrkva[233]).

*Double-layer capacitance measurement.* Formaro and Trasatti[90] detect organic pollutants in concentrations of more than 0.03 mg/l by adsorption on platinum surfaces and capacitance measurement of the electrical double-layer.

*Polarography:* Polarographic water analyses have already been applied by Schwarz[281] for the determination of surface-active water pollutants. They are based on the observation that the oxygen maximum of an oxygen-saturated water mixed with a stock solution of 0.01 N KCl solution decreases as a function of its pollution with surface-active or colloidally dissolved materials when compared to highly purified water.

Breyer and Bauer[37] in particular described the generation of waves in ac-polarography by non-reducible surface-active substances, such as fatty acids, alcohols, phenols, surfactants and organic colorants in aqueous solution (tensammetry).

*Ion exchange:* Riley and Taylor[259] describe the analytical concentration and detection of traces of organic substances, particularly humins from sea water, on Amberlite XAD-1.

*Gas chromatography* with preceding pyrolysis: an overall indication of volatile and nonvolatile carbon compounds can be obtained by gas chromatography of the cracking gases following pyrolysis of the water pollutants. After a water sample is passed through a pyrolysis tube with steam as a carrier and injected into a gas chromatograph (Aerograph 600 C, 3 m chromatographic column packed with Porapak Q at 120°), it is possible to detect less than 10 ppm of mineral oil in 0.15 ml water sample in a flame ionization detector with a standard deviation of ± 1.7 ppm according to Lysyi, Nelson and Webb[211–213].

*Mass spectrometry:* Melpolder, Warfield and Headington[219] have described the determination of volatile water pollutants (particularly hydrocarbons) in the range of 0.01 ppm by mass spectrometry after head space distillation in a $H_2$ stream, dehydration with KOH and condensation.

# 6. CHROMATOGRAPHIC METHODS

## 6.1 Summary

In the enrichment of organic pollutants from water and waste water samples, for example, by extraction or adsorption, mixtures are often obtained with components which often have very different harmful effects in spite of similar physical properties and therefore must be separated into further groups of materials or their individual constituents. Chromatographic techniques often prove to be useful for this purpose; they are characterized by an extremely high resolution and low sample requirements. Some versions were initially characterized by the simplicity of the equipment; however, the increased demands made on savings in labor and time, reproducibility and resolution have often led to high-efficiency instruments now manufactured and marketed by numerous firms.

The common characteristic of chromatographic techniques is the selective separation of a mixture of substances between a stationary and a mobile phase. In *gas chromatography* the mobile phase is a carrier gas in which the vaporized sample is introduced and conducted over the stationary phase. The latter may be a dry adsorbent (gas-solid chromatography) or a separating liquid adsorbed on a solid support (gas-liquid chromatography).

*Liquid chromatography* is involved when the mobile phase is a solvent into which the dissolved sample has been introduced. Again the stationary phase can be a solid adsorbent or a separating liquid incorporated in a solid support (liquid-liquid chromatography). The stationary phase in liquid chromatography takes the form of a column in a tube or a surface; in the case of *paper chromatography* the latter consists of suitable filter paper, while in the more recent *thin-layer chromatography* it consists of other materials with a specific adsorptivity. In *ion exchange chromatography,* the stationary phase consists of a suitable ion exchanger and in *gel chromatography* of an appropriately prepared gel.

Methods for all of the cited chromatographic techniques exist for water and waste water analysis. If the organic pollutants to be analyzed can be vaporized without decomposition, gas chromatography proves to be of advantage because of its sensitivity and resolving power; however, in most cases, thin-layer techniques exist which, although not being as efficient, require somewhat less costly equipment. Liquid chromatography in columns is often used for a preliminary separation of complex mixtures into groups because of the simple design of equipment. Recently, however, refinements in equipment have been reported (high-efficiency column chromatography) which offer similar advantages for non-volatile samples as does gas chromatography for volatile materials.

### 6.2    Thin-layer Chromatography

In thin-layer chromatography the stationary phase is an open surface instead of a column. It was described for the first time by Shraiber in USSR, but became significant in Germany only as a result of the publications of Stahl. It is worthy of note that paper chromatography, which is actually a special case of thin-layer chromatography, had been developed prior to the latter in Western countries and found numerous important applications, including water analysis. At the present time, thin-layer chromatographic techniques are generally preferred to older paper chromatography because as a rule they are more rapid and simple in procedure, with a higher resolving power and more universal application.

In the standard procedure of thin-layer chromatographic analysis, a small quantity of samples (50 $\mu$g — 1 mg) in concentrated solution is applied in spots or bands on the lower edge of a glass plate (about 20 x 20 cm) or film coated with a suitable adsorbent of 0.25 — 0.5 mm thickness and immersed into a flat tray with a suitable solvent system in a closed separating chamber. The solvent rises to the upper edge within 20-60 min and entrains the individual components up to different but specific levels, characteristic for a given substance, in the form of more or less well-defined small eliptical spots as a function of the distribution ratio between solid and liquid phase. The individual spots are detected with suitable reagents or by UV-irradiation and are analyzed quantitatively.

The most common technique today was essentially developed and publicized by Stahl. Recently new auxiliary equipment (for example, that of Desaga/Heidelberg and Camag/Muttenz,

Switzerland) as well as preparations and chemicals (for example, those of Merck) have greatly facilitated the experimental procedure.

Novelties which can also find application in the field of water analysis are the TAS furnace according to Stahl (Desaga). The volatile components of a mixture, for example, an extract, can be directly distilled onto a thin-layer plate in the form of sharp starting points. In a similar device (Diochrom of Camag Company) gas-chromatographic fractions can be transferred onto thin-layer plates.

### 6.2.1  Preparation of Thin Layers

Coating of suitable plate glass slides can be carried out in the laboratory from adsorbents wetted with water and subsequent drying with simple devices. However, Merck Company is marketing ready-to-use coated-glass plates with improved bond strength of the adsorbent as well as ready-to-use coated aluminum foils or plastic films. The latter have the advantage that they can be added to the analytical report as documentation. In the case of glass plates the thin layer which has been developed and analyzed can be stripped by means of an adhesive emulsion (Neatan Merck).

Numerous adsorbents for special purposes are available on the market. The most common are silica gel with 10% gypsum as a binder (Silica Gel G) or alumina G (basic, neutral or acid). Occasionally an inorganic fluorescent agent is added (silica gel $GF_{254}$) which fluoresces after excitations with short-wave UV light. Substances which absorb UV radiation are detected in the form of dark spots without requiring spraying with coloring reagents. Occasionally thin layers are prepared from polyamide (for example, for phenol analysis), kieselguhr, ion exchangers or cellulose. The latter has the advantage of a photo developing time and more well-defined spots when compared to filter paper in paper chromatography.

### 6.2.2  Liquid Phases (Solvent Systems)

For normal water analysis, sufficient experience and information exists with regard to the mixtures to be used as solvent systems for the various groups of materials under consideration (see application examples in the special cards), so that experiments for their selection are usually superfluous. All solvents must be used in highly purified form since impurities can become enriched in the course of the procedure and act as strong interferences.

### 6.2.3    Qualitative and Quantitative Analysis

Various methods are available for the detection of the separated substance spots. Many substances which are colorless in visible light fluoresce when irradiated by UV radiation from an analytical quartz lamp. Inversely, materials which absorb UV-radiation form black spots when they are imbedded in a fluorescent support, as mentioned above. Spraying with concentrated sulfuric acid, which results in black spots, is generally applicable although not specific. Depending on the group of substances involved, specific sprays are available as color-forming reagents, some of which are marketed in the form of aerosols in spray cans.

The development distances of the respective spots, which is substance-specific under fixed operating conditions, serve for a qualitative identification. It is expressed numerically by the Rf-value. This is the development distance of the substance (from the start up to the center of the spot) divided by the difference of the solvent from the spot up to the solvent front when the separating procedure is stopped. Parallel experiments with known substances on the plate serve for a comparison.

Quantitative information can be obtained approximately on the basis of the spot size, particularly in a comparison with the spot size of the same substance of known concentration in parallel experiments. In rough approximation, a logarithmic ratio exists between the area of the spot and the quantity of substance, *i.e.*, a spot of twice the size corresponding to ten times the quantity of substance.

Densitometric comparisons of the color intensity of a spot are considerably more accurate. For this purpose the attenuation of a light beam transmitted through the spot is measured photometrically while the light source is passed transversely through the spot. When the light attenuation is recorded, a bell curve is obtained, the area of which serves as a measure of the quantity of substance present. Suitable calibration curves can be constructed with test solutions of known concentration. In the case of opaque supports (for example, aluminum foils) the readings must be taken by the reflection mode. In a similar manner, fluorescense or fluorescense quenching can be evaluated quantitatively with more recent instruments (for example, from the Camag Company). The Zeiss PMQ II spectrophotometer can also be used.

Another, although complicated, method consists of applying a color-forming reagent on the thin layer, scraping off the spot, extraction and absorbence measurement of the solution obtained.

## 6.3    High-Efficiency Column Coated Chromatography with Continuous Flow Analysis

While column chromatography, as the oldest form of chromatographic methods has brought the most plausible analytical findings in the field of organic analysis for many years in spite of many difficulties, the instrument industry has only recently displayed new interest in the realization of the specifications and success of gas chromatography, as in the columns of liquid chromatography.

Desirable objectives consisted of shortening the separating time, increasing the resolution and continuous flow analysis for a rapid definition and quantitative analysis of the fractions. Suitable measures for the purpose were recognized to be an extension and narrowing of the columns, refinement and standardization of the stationary phase in terms of particle size and, consequently, the need for the use of high pressures to accelerate the mobile phase. As a result, liquid chromatography can have similar advantages of speed and resolving power as gas chromatography even with nonvolatile or thermally unstable substances. The instrument companies are expecting a similar stormy development in the application of their refined products as occurred for gas chromatography in the last 15 years. The advantages are obtained by a considerable increase in equipment cost.

High-efficiency column chromatography is applicable for all of the initially mentioned phase systems (liquid-liquid, liquid-solid, ion exchange and gel chromatography) (see also Ecker[71] as well as Perry[246a]).

A modern liquid chromatograph available from various companies consists of three components (see, for example, Fig. 6.3): the transport system, separating column and detector.

The transport system transports the mobile phase from a reservoir through the separating column under a pressure (up to 300 atm) generated by a suitable pump or from a compressed gas bottle. The solvent is previously passed through a preliminary column packed with the stationary phase utilized, where it is saturated with the solvent system bound to the stationary phase, so that exhaustion of the separation column is avoided. The mixture of substances dissolved in a suitable solvent is injected under

Fig. 6.3—High-efficiency liquid chromatography (Siemens Company).

pressure with a pressure syringe and subsequently transported through the column. The solution eluted from the column is continuously measured in the detector and recorded, and the eluate is passed into a fraction collector for further analyses if necessary.

The separation column consists of a pressure-resistant material—glass for up to 50 atm and stainless steel for higher pressures or, if necessary, the particularly corrosion-resistant tantalum. Column length of about 50 cm and more and inside diameters of 2-3 mm are used. The stationary phase consists of an adsorbent of fine uniform granulometry (in the μm range) and is selected as a function of the method used. Kieselguhr, silica gel, $Al_2O_3$ or special ion exchangers based on polystyrene are used. Occasionally the so-called "controlled surface porosity supports," consisting of a hard glass core with a thin porous surface are recommended. In liquid-liquid chromatography the stationary phase is impregnated with a solvent system having a minimum solubility in the mobile phase, for example, tris-(cyanoethoxy) propane, polyethyleneglycol or the like, similar to gas-liquid chromatography.

The detector is selected as a function of the required sensitivity, the solvent utilized and the most advantageous parameter. Commonly used instruments are (1) the differential refractometer, (2) the photometer detector, (3) adsorption calorimeter, (4) the pyrolysis detector, and (5) for aqueous solutions, the Axt continuous automatic carbon analyzer.

The differential refractometer measures the variation (usually increase) of the refractive index of a solvent as a result of the presence of an eluted substance. An instrument manufactured for this purpose, the Varian Aerograph, still indicates differences of ± $10^{-7}$ units of the refractive index or of 3 $\mu g$ saccharose in aqueous solution. The greater the difference in refractive indices between substance and solvent, the smaller are the quantities of substance which can be determined. A very careful temperature stabilization and pure solvents are prerequisites.

The photometric detector measures the light absorption at a wave length characteristic for the respective substances. For example, in the Zeiss PM4CHR photometric detector, limit concentrations from 0.1 ppm to 0.01 ppb can be determined depending on the light adsorptivity of the respective substance.

In the adsorption count-calorimetry detector (for example, the Varian Aerograph), the heat of adsorption generated by the eluted substances on suitable adsorbents into which thermistors have been imbedded serve as the measured value.

In the pyrolysis detector the substances eluted from the column can be determined continuously in a gas-chromatographic flame ionization detector after evaporation of the solvent and pyrolysis.

Axt[14] proposed the continuous evaporation and combustion of aqueous eluate solutions and determination of the formed $CO_2$ by continuous IR spectrometry.

For literature references to other possible detection methods, see Ecker[71] as well as Perry[246a].

As examples of the applicability of high-efficiency column chromatography in the field of water analysis, two illustrations 9.8 and 14.7 from the Siemens brochure for their S 200 P instrument as well as of the Waters Company are reproduced.

## 6.4 Gas Chromatographic Methods in Water Analysis

As mentioned above, a gas-chromatographic procedure is characterized by the fact that the samples to be analyzed are introduced in vapor form into the stream of a carrier gas and are transported together with the latter into suitable separation columns over a stationary adsorbent surface (gas-solid chromatography, GSC) or over nonvolatile liquids which are applied on solid surfaces (gas-liquid chromatography GLC). The individual components of the sample mixture move through the column at different rates corresponding to their different distribution coefficients between stationary and moving phase, are eluted at the end of the column in separate fractions as individual components mixed with the carrier gas and can then be analyzed quantitatively by means of a suitable detector. The resolving power of the method (even several 100,000 "theoretical plates") and the high sensitivity (limit of detection to $1.10^{-13}$ g) must be emphasized.

Wherever efficient gas-chromatographic instruments and the necessary experience with them are available, these methods will be used for water analysis when a separation into individual components is necessary in vaporizable samples. In many cases the high sensitivity of the method allows the direct injection of water samples without previous enrichment of the organic components to be analyzed. In the contrary case, a larger water sample is extracted with a small quantity of a suitable solvent and the extract is subjected to gas chromatography.

It is assumed in the following that the general principles and applications of gas chromatography are known.

A disadvantage of the method in the analysis of multi-component mixtures of unknown qualitative composition is the difficulty of assigning the detector signals to the individual components on the basis of the measured retention times, so that a second identification technique, for example, IR analysis, mass spectrometry or thin-layer chromatography with a suitable spot analysis must often be used subsequently. Frequently, however, it suffices to combine the indications of specific detectors. As a rule, careful calibration procedures with test mixtures under the given experimental conditions are necessary for quantitative analysis; the total time and labor involved in the first individual analyses are considerable but decrease proportionately in routine analyses.

Suitable and well-equipped instruments are offered by numerous companies, admittedly at formidable prices. The manufacturers also offer tested instructions for common applications, for example, for the determination of pesticides, hydrocarbons, phenols and the like.

### 6.4.1   Gas Chromatography Equipment

Many companies manufacture gas chromatographic equipment in the most diverse designs. Depending on price, they are equipped with the most diverse accessories in order to increase convenience, accuracy, resolution and range of application. The support materials, as a rule based on diatomaceous earth, are available in purified form and in different granulometries. The end designation AW (acid wash) refers to products purified with acids, while the code DMCS (dimethylchlorsilane) or HMDS (hexamethyldilazane) refers to silanized products which eliminate undesirable adsorption effects (tailing of peaks). For low molecular weight, highly volatile substances the Porapak product has proved to be useful; it consists of polystyrene cross-linked with divinylbenzene and can be used without a stationary separating liquid. A large selection of stationary separating liquids also exists. Unfortunately they are frequently identified by code names by the manufacturers, so that the results obtained with them are not always applicable for other products. In Germany, many well-proved supports and other accessories for gas chromatography can be obtained from the Merck Company in Darmstadt.

### 6.4.2   Detectors

A choice of the most suitable detector is important for application in water analysis. Because of its universal applicability—responding to all organics to be considered—and its high sensitivity, the flame ionization detector (FID) is used most frequently.

It is advantageous to use selective detectors, in addition. Among these, the electron capture detector, (ECD) responds preferentially to halogenated hydrocarbons and is therefore particularly suited for the determination of minimal quantities of chlorine-containing insecticides and weed killers. The necessary current of slow electrons is delivered either by weak $\beta$-emitters (tritium or $^{63}$Ni) or by a strong arc in a helium stream.

A detector which preferentially detects phosphorus compounds is the alkali FID. A dish with rubidium sulfate or another alkali salt is introduced into the flame zone. While maintaining suitable operating conditions, particularly the hydrogen feed, organophosphorus compounds produce an approximately 10,000 times stronger response than phosphorus-free substances.

For selective gas chromatography it is of advantage to use an instrument with a double-column configuration, one column being equipped with the universal FID and the other with the selective detector. By comparison of the two chromatograms it can be determined which peak of the FID corresponds to a halogen- or phosphorus-containing substance. This considerably facilitates quantitative analysis.

Mechanical or electronic integrators are used increasingly for the calibration and quantitative evaluation of the chromatograms obtained.

Examples for the use of gas chromatography for phenols, hydrocarbons and pesticides are described in the special part of the book.

*The coulometry detector.* A special degree of specificity is obtained with the use of the coulometry detector. It was proposed for the first time by Coulson, Cavanagh, De Vries and Walter[55] (see also Coulson and Cavanagh[54] for the selective gas chromatography of chlorine- and sulfur-containing compounds. In recent years the Dohrmann Instrument Company (Mountain, California) has marketed suitable instruments which permit a considerable refinement of techniques (see, for example, Burchfield and Wheeler[44a], Guiffrida and Ives[106] as well as the application instructions of the cited firm).

For the coulometric detection of diluted gas-chromatographic fractions, the latter are conducted through fine reaction tubes in which they are either subjected to catalytic combustion at 800° with an oxygen addition with the formation of HCl or $SO_2$ or are reduced in a hydrogen stream and decomposed with quantitative formation of $PH_3$ or $NH_3$. The reaction gases are conducted into a coulometric microtitrator (see Fig. 6.4.2), where they are

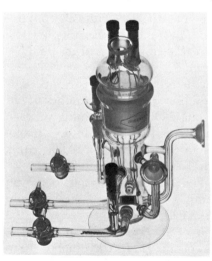

Fig. 6.4.2—Coulometry detector (Dohrmann Company).

absorbed and subjected to the appropriate quantitative titration reactions. The hydrogen chloride which is absorbed in 75% acetic acid is precipitated with silver ion and the $Ag^+$ concentration of the solution measured on a silver electrode is reconstituted from a silver anode by electrolysis. The current utilized for this purpose is recorded as peaks on a recorder as customary in gas chromatography. This method permits the determination of chlorine from chlorinated hydrocarbons even in the nanogram range and with special devices even at lower concentrations.

Sulfur from sulfur compounds of all types is converted into $SO_2$ during combustion absorbed in dilute acetic acid in an electrolyte from KI and is automatically titrated with iodine which is formed by electrolysis. A platinum wire serves as the indicator electrode.

The reduction of phosphorus compounds into $PH_3$ takes place in a hydrogen stream at 950° in a quartz tube. Like HCl, $PH_3$ is determined coulometrically with silver solution. Hydrogen chloride or hydrogen sulfide forming simultaneously is removed previously over aluminum oxide. Recently, the detection of nitrogen compounds has been described by quantitative determination of the $NH_3$ formed during reduction in a $H_2$ stream on nickel catalysts, allowing the nitrogen to be detected in quantities of only 0.5 ng.

The use of the coulometry detector for a selective determination of compounds containing Cl, S, P or N has further advantages in addition to specificity. Since compounds which do not contain the element to be determined (for example, C, H and O) are not recorded, all interferences which may be expected in other detectors by the solvent, by volatilizing separation phases as well as incomplete separation of byproducts remains absent. A quantitative analysis of the peaks is also considerably simpler; it takes place by stoichiometric calculation instead of on the basis of individual factors, which differ greatly, for example, in the electronic capture detector.

### 6.4.3  Head Space Analysis

The analyst frequently has the task to determine the volatile organic pollutants responsible for the odor of a water or waste water sample separately from nonvolatile components. Head space analysis is suitable for this purpose. The Multifract F 40 has been developed for it by the Perkin Elmer Company. The water sample is brought to equilibrium with the air space in a closed vessel at constant temperature. The corresponding volume of gas phase is then transferred into a customary separation column and analyzed. Quantitative analysis requires model experiments with the same volume ratios.

This permits the determination of 1 ppm petroleum ether in water. The volatile components of waste water, for example from an alcohol distillery, can be determined in a similar manner without interference from higher-boiling components.

*Head space analysis in the autoclave.* Gjavotchanoff, Lüssem and Schlimme[107] enrich volatile organic components in the head space in an autoclave which is filled to about two-thirds with the water sample while the balance remains empty. The autoclave is heated to 160-180° with the development of 5-15 atm (rel.) pressures. Two ml of the gas phase are removed with a sealed syringe, injected into a gas chromatograph and analyzed under customary experimental conditions.

### 6.4.4  *Reverse-Gas Chromatography*

If volatile components are present in a water sample in such small quantities that they require an enrichment, this has recently become possible successfully by outgassing with a carrier gas, en-

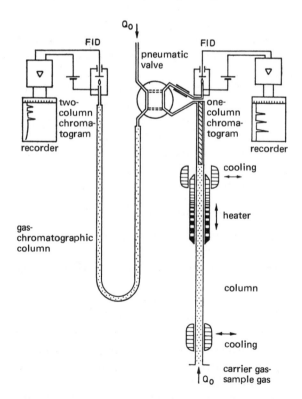

Fig. 6.4.4a—Flow sheet of the reversed gas chromatograph with optional series-connected gas-chromatographic column.

richment and separation by reversed gas chromatography (Kaiser [155-156], Oster[244], see also Leithe[196]).

Enrichment takes place in a tube packed with a suitable adsorbent. The two ends of the packing can be cooled with cooling devices and a heating sleeve can be moved along the entire tube.

The process operates in a batch mode. In the first period the carrier gas containing the volatile component is passed through under the full action of the two cooling zones, thus releasing the adsorbable components at the tube inlet. In the next step the heating sleeve is passed over the entire tube at the selected desorption temperature. This results in fractional desorption up to just before the tube ends. Finally, the heating sleeve is passed over the cold zone at the tube end; the completely desorbed and simultaneously preseparated fractions are finally separated completely in a following standard gas chromatograph and measured in a suitable detector (Figs. 6.4.4 a and b). In the meantime the cold zone at the tube inlet has been cooled again, a new gas

Fig. 6.4.4b—Reverse gas chromatograph apparatus of Siemens Company.

sample has released the adsorbable components there and in its further path assumes the function of an effective gas chromatographic carrier gas.

According to Kaiser, $5.10^{-10}$ vol.% butane can be determined in nitrogen, for example, by this method.

Kaiser[156] has developed a minivariant of reversed gas chromatography for the enrichment and detection of volatile organic trace elements in water. The volatile components are driven from the water sample by the introduction of 2-16 l air or another carrier gas and for adsorption are passed through a quartz or steel tube packed with 0.5 g adsorbent (5% Dexsil 300 on Chromosorb A.) Previously a temperature gadient of $-160°$ to $+20°$ was established in the adsorption tube by the introduction of nitrogen or air, cooled by flowing into liquid air or liquid nitrogen. Desorption takes place in a hot air stream (400°-150°) in a gas chromatograph. Even 0.1 ppb of volatile organic trace elements can be detected in 10-200 ml water by a flame ionization detector.

# 7. VOLATILE FATTY ACIDS

Volatile fatty acids occur in residential waste water, in agricultural silage waste effluents as well as in many industrial effluents, particularly from processing of natural products but also from petrochemical synthesis. Their presence frequently becomes evident by their odor after acidification or gentle heating.

In the presence of larger quantities, they can be determined by customary acidimetric titration in the condensate after distillation from the water sample which has been acidified with $H_2SO_4$ (see DEV H 21). If necessary, the water sample, adjusted to a pH of about 10, can be previously enriched by evaporation.

Colorimetric methods make use of the red color of hydroxamic acid, prepared with hydroxylamine, after addition of $FeCl_3$.

The most favorable method is to use gas chromatography, since this also allows a separation into the individual homologous acids.

The higher fatty acids (above $C_{10}$) are present in sludge as a result of formation of sparingly soluble calcium and magnesium soaps. They can be readily obtained after acidification by ether extraction.

The determination of short-chain fatty acids by direct injection into the gas chromatograph has been described by van Huygsten[144]. Two $\mu$l of the sample, acidified to pH $1-2$ with HCl, are loaded on the separation column (2 m length, 3 mm I.D. with 3% FFAP [= free fatty acid phase from polyethyleneglycol and 2-nitroterephthalic acid] on Chromosorb 101—a styrene-divinyl-benzene copolymer). In a Beckman GC-4 with a double FID, temperatures of 240° are adjusted at the inlet, 180° in the furnace and 330° in the detector. Gas flow rate 77 ml/min $N_2$, 45 ml/min $H_2$, 280 ml/min air.

The colorimetric determination on iron (III) hydroxamates has been described by Montgomery, Dymock and Thom[227]; see Harwood and Huyser[126] as well as Grütz[117] and Wagner[321a] for the automatic procedure in an autoanalyzer.

# 8. CYANIDES

## 8.1 Introduction

Because of their toxic action, cyanides are important components of trade and industrial wastes effluents. They are absent from domestic waste water.

In industrial waste effluents the cyanide pollution originates more frequently from galvanizing plants; larger quantities, although from fewer places, occur in the gas liquor from coking plants, in blast gas purification of blast furnace installations, and in electro-chemical production processes, for example, with the use of Söderberg electrodes for aluminum production. Large quantities of hydrocyanic acid and cyanides are employed industrially for ore leaching for noble metals, flotation agents for zinc and lead ores, surface treatment of steels, post-pickling of metal parts as well as in pesticides. In organic synthesis hydrocyanic acid serves for the production of nylon and Plexiglas, among other things, and it is a waste product in the manufacture of acrylonitrile from propylene and ammonia.

Earlier, coking plants were more widespread than today since many smaller towns also obtained their city gas supplies by coking of coal in their own plants. Today most cities have converted from gas generation to natural gas which is supplied to homes either directly or after partial conversion into hydrogen and carbon monoxide, or a cracking gas is produced from bottled gas or light naphtha. A considerable relief of the sewers has occurred in many places as a result.

*Cyanide chemistry.* Cyanides occur in water in the following forms:

1. Free hydrocyanic acid. As a very weak acid (dissociation constant $7.2 \cdot 10^{-10}$) it is released from its alkali salts already by carbon dioxide and, because of its volatility (b.p. 25.7°) it can be blown out from aqueous solution by carrier gases at room temperature. The distribution constant k (mg HCN/l carrier gas/mg HCN/l water) $= 3 \cdot 10^{-3}$ (Schneider and Freurd[276]).

2. Simple cyanides (alkali and alkaline earths cyanides).

3. Easily decomposable complex cyanides (for example, of Zn).

4. Sparingly decomposable complex cyanides: $[Fe^{III} (CN)_6]$, $[Fe^{II} (CN)_6]$, $[Co(CN)_4]$.

Complex nickel and copper cyanides are assuming an intermediate position between 3 and 4.

*Toxicity of cyanides.* The toxicity of cyanide-containing water depends on the state of the cyanide in aqueous solution. For a fish it is toxic particularly as free hydrocyanic acid; as mentioned above, its concentration in cyanide solution is a function of the pH value. According to Meinck, 0.02 mg CN/l has no effect on fish, while 0.05 mg is lethal for trout in 5 days. In *Daphnia,* the minimum toxic growth is 0.01 mg/l. In the case of biological treatment without adaptation, the toxic effect begins at 0.65 mg/l, while adapted bacteria can still degrade 30 mg/l. The critical toxic concentrations of complex cyanides are listed in Table 8.1 according to Bucksteeg.

Table 8.1. Critical concentrations in mg CN/l.

|  | Fish | Daphnia |
|---|---|---|
| $K_3Cu(CN)_4$ | 1.0 | 0.8 |
| $K_2Ni(CN)_4$ | 30.0 | 75.0 |
| $K_2Zn(CN)_4$ | 0.3 | 13.5 |
| $K_2Cd(CN)_4$ | 0.75 | 0.5 |

For man HCN and free $CN^-$ are the principal toxic forms. They are formed particularly from alkali and alkaline earth cyanides as well as from the easily splittable complex cyanides (of Zn, Cd). Complex cyanides which are not split even by gastric juice are considered to be of low toxic or nontoxic $[Fe^{II} (CN)_6, Fe^{III} (CN)_6]$. For man the lethal HCN dose amounts to 60 mg.

The toxicity of complex cyanides can increase in the further half traveled by waste water, for example, by the entry of acid or by irradiation.

## 8.2   Sample Preparation

In order to avoid hydrocyanic acid losses during storage, cyanide containing water samples are brought to pH 11 by the addition of NaOH.

In order to remove $H_2S$, which is volatilized in the same manner as HCN during distillation, the sample, adjusted to pH 11, is shaken with a small amount of powdered lead carbonate and filtered. If distillation from tartaric acid solution is planned, a few drops of 10% $CdSO_4$ solution are added. Since CdS is insoluble in tartaric acid, filtration is unnecessary.

*Separation of cyanides.* The light HCN liquid allows simple isolation by distillation from mixtures containing interfering impurities. The choice of distillation conditions depends on the purpose of the analysis, whether the total cyanides or only the highly toxic cyanides without the relatively stable and therefore nontoxic cyanoferrates are to be determined.

Oxidative impurities, particularly free chlorine, the presence of which can be determined with potassium iodide-starch paper, aer made harmless by the addition of ascorbic acid.

Volatile fatty acids which would disturb during alkali titrations by saponification are removed without significant HCN losses at a pH of 6—7 by a single extraction with isooctane or hexane (1/5 of the water volume).

The methods described in the literature essentially differ by the pH value of the acidified mixture during distillation. In order to liberate the total HCN from complex cyanides, distillation is carried out from inorganic acid—as a rule sulfuric acid—solution, although an excessively large acid addition and prolonged boiling times should be avoided because of the risk of cyanide losses by saponification into $(NH_4)_2SO_4$. Furthermore, various additives are described which facilitate HCN separation with the formation of new more stable complexes (EDTA, KI, Cu(I)-, Hg- and Mg-salts). An intermediate alkaline decomposition step is recommended by Bucksteeg and Dietz. Only the complex cobalt cyanide has proved to be practically undecomposable.

Some methods provide for a reflux condenser on the decomposition flask and a weak air stream in order to transfer the liberated hydrocyanic acid into the alkaline receiver, while others distill directly into the receiver up to a certain residue of liquid. In the further treatment of the receiver contents in which hydrocyanic acid is present in enriched form, suitable caution is necessary because of its toxicity.

Occasionally an analytical distinction is made between "chlorine-decomposable" and "undecomposable" cyanides with reference to the customary technical removal of cyanides with chlorine in alkaline solution.

*End determination of the liberated cyanide.*

Recently, selective membrane electrodes have been introduced for potentiometric $CN^-$ determinations.

According to Liebig, larger quantities of cyanide in the distillate can be titrated in the presence of potassium iodide. For quantities for more than 5 mg $CN^-/l$, Fiegl's silver reagent (dimethylbenzalrhodanine) has proved useful as an indicator. Huditz and Flaschka have described a complexometric titration method with nickel salt and murexide as indicator.

For low $CN^-$ concentrations of about 0.20-10 mg/l colorimetric methods via bromine or chlorine cyanide with pyridine and benzidine, antipyrine, barbituric acid, p-phenylenediamine or dimedone are customarily used.

Since the toxicity of the cyanide in water depends primarily on the free hydrocyanic acid concentration, methods in which these are removed from the water sample without disturbing the equilibrium between HCN and $CN^-$ are of interest. This can be realized by extraction with methylchloroform or by blowing off at room temperature and gas chromatography according to Schneider and Freund.

## 8.3   CN-Determination with Selective Membrane Electrodes

The Orion Company, Cambridge, Mass., USA, is manufacturing selective membrane electrodes from a mixture of $Ag_2S$ and AgI for a potentiometric $CN^-$-determination; $10^{-2}$ to $5 \times 10^{-5}$ M $CN^-$ solutions can be analyzed in continuous flow measurements. According to Fleet and Storp[88] they are suitable for water analysis. The anions are interferences only for iodide and sulfide. Only free $CN^-$ ions but not stable $CN^-$ complexes are indicated (see also Blaedel et al.[28]).

## 8.4   Titration Methods

### 8.4.1   Liebig cyanides titration (for CN⁻ concentrations of 10 mg/l) (DEV) D 13)

Of the distillate obtained, for example, according to 8.7, 250 mg are treated with 5 g ammonium sulfate and 0.2 g potassium iodide, and are titrated with 0.1 N $AgNO_3$ solution until the material assumes a milky-yellow turbidity. A blank reagent value is subtracted from the result.

The calculation is based on the following reaction equation:

$$2 \, CN^- + AgNO_3 = [Ag \, (CN)_2]^- + NO_3^-$$
$$1 \text{ ml } 0.1 \text{ n } AgNO_3 = 5.2 \text{ mg } CN^-$$

*Titration with Feigl's silver reagent (more than 5 mg $CN^-/l$)*
*(DEV D 13)*. The distillate (about 250 ml) is treated with 0.5 ml
indicating solution. Titration is carried out with 0.02 N $AgNO_3$
up to the color change from light yellow (canary yellow) to red-
dish yellow (salmon yellow) which was recorded in a blank experiment.

$$1 \text{ ml } 0.01 \text{ N } AgNO_3 = 0.52 \text{ mg } CN^-.$$

Indicator solution (Feigl's silver reagent): 0.02 g 5- (4'-dimethyl-
benzalrhodanine is dissolved in 100 ml acetone. The solution is
stable for about 2 weeks.

### 8.4.2   Complexometric Titration According to Huditz and Flaschka[141]

Complex formation of the cyanide with nickel salts takes place
according to

$$4 \text{ } CN^- + Ni^{++} = Ni(CN)_4^{2-}$$

The addition of excess nickel ion can be recognized with murexide, a complexometric indicator, by the color change from
blue-violet to orange-yellow. Titration is carried out in ammoniacal solution. The presence of halides, rhodanide and complex ion cyanides does not interfere.

Reagents: 0.1 M nickel sulfate (adjusted gravimetrically with
dimethylglyoxime or complexiometrically with Complexon III
against murexide). One ml of solution corresponds to 10.4 mg
$CN^-$. Small quantities of cyanide can also be titrated with 0.01 M
nickel solutions.

Indicator powder: murexide is thoroughly ground with NaCl,
1: 500. Procedure: the sample solution is treated with 30 ml concentrated $NH_3$ per liter (NaOH alkaline solutions are treated
with the corresponding amount of ammonium sulfate). Indicator
powder is added until a strong blue-violet color forms and titration follows immediately with the adjusted nickel solution up to
the change to orange-yellow.

For very hard water samples Pociecha[250a] recommends pretreatment with a strong acid base exchanger which was converted
into the $NH_4^+$ form shortly before use by washing with 10%
ammonium chloride solution and subsequently with water.

## 8.5     Colorimetric Methods

### 8.5.1     Colorimetric Determination with Pyridine-benzidine (0.02-10 mg CN⁻/l) (DEV D 13)

A disadvantage of this method is the carcinogenic action of benzidine. An aliquot of distillate is brought to 50 ml and adjusted to pH 7 by addition of 2 ml phosphate buffer. Three ml bromohydrochloric acid are added, the mixture is shaken, 3 ml arsenite solution are added, followed by thorough mixing. The solution should become colorless. Fifty ml amylalcohol are added, the solution is well shaken, followed by addition of 5 ml pyridine-benzidine mixture and shaken again. The pH should range between 4.3 and 4.5. After standing for 30 min the amylalcohol phase is filtered and after a total of 40-90 min since the addition of the pyridine-benzidine mixture, a photometric reading is taken at 491 nm.

Solutions: arsenite solution: 26 g $As_2O_3$ and 23 g NaOH are dissolved in 700 ml water, adjusted to pH of 4-5 with 0.1 N HCl and diluted to 1 l.

Benzidine solution: 2 g benzidinium chloride and 0.2 ml hydrochloric acid (d = 1.19) per 100 ml water.

Pyridine solution: 25 ml ultrapure pyridine and 2 ml hydrochloric acid (d = 1.19) per 75 ml water.

Pyridine-benzidine mixture: 50 ml pyridine solution and 5 ml benzidine solution.

Phosphate buffer: 25.4 g $Na_2HPO_4 \cdot 2\ H_2O$ and 75.4 g $KH_2PO_4$ are dissolved in 1 l water.

Bromohydrochloric acid: 5 ml bromine are dissolved in 1 l of 2 N HCl.

KCN standard solution (1 ml = 1 mg CN⁻): 1.277 g of KCN are diluted to 500 ml with water. Adjustment on the basis of Liebig titration.

*Colorimetric determination with pyridine-barbituric acid (0.02-2 mg CN) according to Asmus and Garschagen*[13a]. Procedure according to Bucksteeg and Dietz[41]: 10-25 ml of the alkaline distillate or, in the case of concentrations higher than 2 mg CN/l, a 10-fold dilution obtained with 0.2 N NaOH, are adjusted to pH 6.0 with 15 ml phosphate buffer solution. One ml chloramine-T solution is added, followed by mixing, and after 1 min, addition of 3 ml barbituric acid-pyridine solution, dilution to 100 ml (pH of the solution about 6) and photometry after 20 min at 570 nm. The calibration curves are constructed with KCN solutions in 0.2 N NaOH with 0.02-2 mg CN'/l.

Reagents: phosphate buffer, pH = 6.0.

Chloramine T-solution, 1%, in water.

Barbituric acid-pyridine solution: 15 g barbituric acid are suspended in a small amount of water and dissolved after addition of 75 ml ultrapure pyridine. Fifteen ml hydrochloric acid (d = 1.19) are added, followed by cooling and dilution with water to 250 ml. Leschber[200] adjusts the alkaline distillate to pH = 8-9, while Mertens[220] adjusts to pH = 5.4 with 0.1 ml succinic acid buffer.

*Colorimetric determination with antipyrine (ASTM D 2036; see also Epstein[77] as well as Ludzack, Moore and Ruchhoft[208]).* A quantity of distillate containing 10 $\mu$g − 1 mg CN'/l of water sample is adjusted to a pH of 6.5-8.0 with 10% acetic acid and mixed with 0.2 ml chloramine-T solution in a 50 ml graduated flask. After 1-2 min 5 ml pyridine-pyrazolone reagent are added, the solution is brought to the mark with water and after 20 min the blue color is read at 620 nm.

The calibration curve is constructed with a standard solution containing 1 $\mu$g CN$^-$/ml (by 1000-fold dilution of a solution of 2.51 g KCN + 2 g KOH/l). Dilutions of 1-10$\mu$g CN' in 50 ml indicator solution are read.

Reagents: chloramine-T solution, 10 g/l.

Pyridine-pyrazolone reagent, solution A: 0.25 g 1-phenyl-3-methyl-5-pyrazolone is dissolved in 50 ml water of 60° and cooled.

Solution B: 0.01 g bis-pyrazolone in 10 ml pyridine.

Mixed reagent: solution A and B are filtered successively through the same paper filter and mixed. A pale pink color does not interfere.

Bis-pyrazolone is available as number 6969 from Eastman Kodak. Preparation: 17.4 g phenylmethylpyrazolone are dissolved in 100 ml 95% ethanol, mixed with freshly vacuum-distilled phenolhydrazine, and boiled for 5 h with refluxing. The insoluble reaction product is removed by suction and the product is washed with hot alcohol. The melting point is higher than 320° and in dry form the product has indefinite stability.

For the necessary purity of the reagents, see Bark and Higson[19].

*Colorimetric determination with pyridine-p-phenylenediamine.* According to Bark and Higson[19] (see also Montgomery, Gariner and Gregory[226a]), 10 ml neutral distillate are mixed with 0.2 ml saturated bromine water in a 20 ml graduate flask; after 5 min 0.4 ml 2% arsenous acid is added as a decolorizing agent, followed by 8 ml reagent mixture and bringing to the mark. After 30-50 min, the photometry reading is taken at 508 nm. Reagent mixture: 3 ml pyridine solution (150 ml pyridine + 100 ml water + 25 ml conc. HCl) are mixed with 1 ml p-phenylenediamine solution (1.7 g p-phenylenediamine dihydrochloride in 500 ml 0.5 N HCl).

For the automatic procedure of this method in the autoanalyzer (40 samples/hour) see Casapieri, Scott and Simpson[47].

*Colorimetric determination with dimedone according to Mirsch*[222]. Twenty-five ml of the cyanide-containing distillate collected in 0.01 N NaOH are treated with 1 ml 1% chloramine-T solution; after 1 min of standing, 3 ml dimedone reagent (3% dimedone in 30% pyridine) and 10 ml phosphate buffer, pH 7.6, are added, diluted to 50 ml and after standing for 45 min, read on the photometer at 585 nm.

## 8.6 Total Cyanide Determination According to Bucksteeg and Dietz[41]

The method of Bucksteeg and Dietz, described below, is essentially based on ASTM method D 2036 which was supplemented by intermediate alkali treatment.

**𝕊** = standard taper joint

Procedure: the receiver in the distillation equipment (Fig. 8.6) is charged with 20 ml N NaOH. An air stream is aspirated through the system at 5-10 bubbles/sec, the flask is charged with 10 ml EDTA solution and 100 ml of the water sample, if necessary freed from oxidants (especially free chlorine) with ascorbic acid at pH = 11. Moreover 10 ml $HgCl_2$ solution are added. After addition of 5 drops of methyl orange, the solution is neutralized with 2 N HCl and acidified with 10 ml of this acid. The flask is heated and boiled for 30 min in the above air stream. Without interrupting the distillation, 10 ml $MgCl_2$ solution and 15 ml 4 N NaOH solution as well as 20 ml $HgCl_2$ solution are added, followed again by boiling for 20 min. Finally 30 ml $H_2SO_4$ (1 + 1 vol.) are added in small portions, followed by distillation for 1½ h. The alkaline distillate is brought to the mark in a 100 ml graduated flask. Aliquots are determined either by volumetric analysis or colorimetry.

Reagents: ascorbic acid solution, 2 g in 100 ml water.

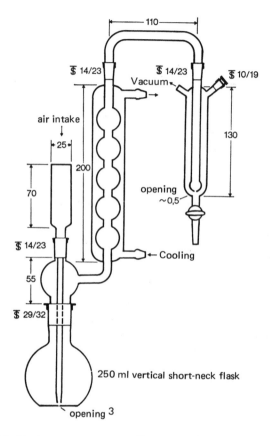

Fig. 8.6—Distillation apparatus according to Bucksteeg and Dietz.

EDTA solution: 10 g disodium ethylenediaminetetraacetic acid and 90 ml water.

HgCl$_2$ solution: 68 g HgCl$_2$ and 1 l water.

MgCl$_2$ solution: 510 g MgCl$_2$·6 H$_2$O and 1 l water.

## 8.7    Partial Cyanide Determination

*Cyanide distillation with tartaric acid according to DEV.* DEV D 13 prescribe distillation of the cyanides in tartaric acid solution essentially on the basis of the instructions of Ludzak, Moore, and Ruchhoft[208]. Only the free and easily decomposed cyanides are to be determined, while the stable ion cyanides are left undecomposed. Several authors (see, for example, Bucksteeg and Dietz[41]), however, reported unsatisfactory results. In any case, an addition of 1 g $Pb(NO_3)_2$ or cadmium acetate (Leschber[200]) seems to be advisable. See also Gad and Schlichting[100] as well as Mertens[220] who adjusts to pH = 5.6 with a succinic acid-NaOH buffer.

For the complete prevention of decomposition of the stable complex cyanides of lower toxicity, Boye[34] distills at pH = 7.2-7.4 after addition of diethylbarbituric acid (Veronal), while Fleps[89] recommends a pH of 7.8-8.8.

With regard to the separation of HCN under an air stream at pH = 7, see Jenkins, Hey and Cooper[151]; see also Roberts and Jackson[260] as well as Russell and Wilkinson[266] concerning vacuum distillation at 127-254 mm Hg.

Procedure according to DEV D 13: if the water sample was alkalinized with NaOH at the time of sampling for preservation, 100 ml of sample are first titrated with tartaric acid solution against methyl orange to the end point of a red color. Subsequently 250 ml of water sample are first treated with 2.5 times the volume of tartaric acid necessary to acidify the 100 ml sample and then with an excess of 5 ml. About 250 ml are distilled slowly and uniformly in distillation equipment with a descending condenser and an adapter immersed in 50 ml NaOH.

See also Bahensky and Zika[15] on distillation results at different pH values.

*Determination of "decomposable" cyanides according to Bucksteeg and Dietz*[41]. According to Bucksteeg and Dietz the cyanides which can be decomposed by chlorine or catalytically (alkali cyanides, cyanide complexes of zinc, cadmium, copper, silver and nickel) but not the iron and cobalt cyanides are determined by the method described below. Distillation is carried out at pH = 5.0-5.2.

Procedure: the consumption of N HCl for neutralization is first determined against methyl orange in a 100 ml sample. In the distillation equipment described in 8.6 and shown in Fig. 8.6, the washer is first charged with 20 ml N NaOH and the air flow is adjusted to 5.10 bubbles/sec. The flask is charged with 10 ml zinc acetate solution, 10 ml lead acetate solution, 25 ml buffer solution and 100 ml water sample. The volume of N HCl consumed in the preliminary experiment is then added, followed by heating to boiling for 3/4 h in the air stream with refluxing. The suspenpension of 0.2 g zinc dust is then added without interrupting the distillation and distillation is again performed for 3/4 h. The distillation is then interrupted, the alkaline distillate is transferred into a 100 ml graduated flask and the cyanide is determined as described below.

This treatment also attacks a part of the complex iron cyanides (their quantity is obtained from the difference of total cyanides "decomposable" cyanides). It can be determined on the basis of a correction curve (see the original study).

Zinc acetate solution: 100 g $Zn(CH_3COO)_2 \cdot 2\ H_2O$ in 250 ml water.

Lead acetate solution: 120 g $Pb(CH_3COO)_2 \cdot 3\ H_{20}$ in 1 l water.

Buffer solution: dissolve 200 g NaOH + 240 g citric acid each in 400 ml water, combine the solutions with cooling and add 170 ml glacial acetic acid.

*Determination of NaOCl-decomposable cyanides according to ASTM.* An aliquot of water sample is adjusted to pH = 11-12 with NaOH. A solution containing 50 g $Ca(OCl)_2/l$ is added in drops until a distinct blue color appears on potassium iodide starch. The excess chlorine indicated here is maintained while the solution is stirred for 1 h. The chlorine excess is then reduced by addition of 0.5 g ascorbic acid, an excess of 0.5 g is added for the sake of safety, and the CN content is determined in the untreated as well as in the chlorine-treated water sample. The difference yields the Cl-decomposable fraction.

## 8.8 Determination of Free Hydrocyanic Acid by Extraction with Methylchloroform

Since the acute toxicity of cyanides for water organisms depends essentially on the content of free undissociated hydrocyanic acid, Montgomery, Gardiner and Gregory[226a] determine this component of a water sample without shifting the equilibrium by changing the dissociation conditions. They extract 100 ml of water sample with 35 ml methylchloroform (1,11-trichloroethane) and extract the hydrocyanic acid from the organic phase by shaking with 10 ml of a solution of 2 g $Na_4P_2O_7 \cdot 10\ H_2O$ in 100 ml water. The bound hydrocyanic acid in the aqueous alkaline extract is determined according to Bark and Higson (p. 108).

*Gas-chromatographic determination of free hydrocyanic acid by displacement in an air stream according to Schneider and Freund[276].* The water sample is blown out with an air stream which has been dried on magnesium perchlorate and the free hydrocyanic acid is enriched without disturbing the sample equilibrium in a precolumn of 18 cm length packed with 20% dinonylphthalate on 40-60 mesh milled brick and cooled with acetate-carbon dioxide. Desorption is realized by immersing in water of 57° in a helium flow and the latter is conducted into a Fractometer 154 B (Perkin Elmer) equipped with a 6 m separating column packed with 20% dinonylphthalate on Chromosorb W and equipped with a thermistor detector. Aqueous HCN solution of 30 $\mu$g to 15 mg HCN/l were analyzed.

Claeys and Freund[51] make use of two columns for the gas-chromatographic determination of hydrocyanic acid desorbed from the precolumn. One of these is 2.7 m long and is packed with 15% TECP [1,2,3-tris-(2-cyanoethoxy) propane] on Chromosorb W/DMCS; the second is 1.8 m long and is packed with 50-80 mesh Porapak Q. Temperatures: column 1 = 37°, column 2 = 51°. Carrier gas: nitrogen, 75 ml/min; FID 35 ml $H_2$; 400 ml air/min. The symmetrical HCN peak appears after 5 min.

# 9. PHENOLS

## 9.1 Summary

Phenols, *i.e.*, hydroxy compounds of aromatic hydrocarbons (benzene and homologs, naphthalene and the like) are frequent pollutants of industrial waste water and can therefore also be detected in polluted rivers. They are present primarily in the waste effluents of coking plants and brown coal distilling plants, even if most of them have been removed by extraction according to regulations. Furthermore, phenols are components of important plastics, are raw materials for drugs, dyes and, particularly chlorophenols, are present in weed control agents and other pesticides, making them frequent pollutants of the waste effluents of organic synthesis.

Phenolic compounds furthermore are components of pulp effluents and also form during the decomposition of leaves. Phenols of natural origin are present in detectable concentration in human and animal urine.

While phenol as such can be detected in water by its odor or taste even in concentrations of $0.01 - 0.1$ mg/l, this effect is enhanced by $1 - 2$ orders of magnitude, *i.e.*, to $1$ $\mu$g/l, under the influence of chlorine used for drinking water disinfection as a result of the formation of chlorophenols which have a stronger odor and taste.

Phenols are toxic for most organisms; as such, they are bacterial toxins, but they are increasingly subject to biodegradation by phenol-resistant bacterial strains. It is therefore necessary to poison phenol-containing waste water which cannot be immediately analyzed by alkalinization, acidification or by the addition of $HgCl_2$ or $CuSO_4$.

For the fishing industry phenol is of interest not only because of its toxicity. Sublethal concentrations (beginning with $0.1$ mg/l) become enriched in fish and make it unconsumable if the flavor is not regenerated by maintaining the caught fish in clean water for a fairly long period of time.

Phenol enrichment from water samples can be realized with suitable solvents. The distribution coefficients $K = C_E C_W$, where

$C_E$ and $C_W$ are the phenol concentrations in the extractant and water, respectively, amount to 55 for diethylether, 52 for 2-heptanone, 2.1 for benzene and 1.3 for dichloromethane, according to Dietz and Koppe[63]. At a pH value of 8, the phenols are extracted completely, while only a small amount of the fatty acids is coextracted. The phenols from the organic phase are transferred into the aqueous-alkaline phase with N NaOH and after acidifying with $H_2SO_4$, are extracted twice from this phase with 10 ml ethylether. Evaporation of the ether solution without losses to about 20 $\mu$l can be carried out in the apparatus of Fig. 3.2c after addition of 100 $\mu$l diethylamine.

From the standpoint of analysis, we distinguish volatile phenol (for example, phenol, cresols, xylenol) from non-volatile phenols (di- and trihydroxy compounds). The first group is of greater interest with regard to its influence on taste, and it is for this reason that the analysis is preceded by steam distillation, which also has the purpose of eliminating secondary components. $H_2S$, as an impurity, usually is eliminated by precipitation.

The older analytical methods made preferential use of colorimetry. However, as a rule these do not equally respond to all homologs, and respond not at all or only slightly to p-cresol and dichlorophenol. Since usually only phenol is used for calibration, the data in the presence of higher homologs are not valid for the entire group of compounds and can therefore only be used as guideline values. In Germany, a diazo-reaction with p-nitraniline is customary, while in the US a reaction with dibromoquinone-chlorimide (Gibbs reagent) is occasionally used in addition to the reaction with aminoantipyrine.

Considerably more complete information is obtained from chromatographic separation with a subsequent determination of the individual components. Gas chromatography as well as column, paper and thin-layer chromatography are customary techniques.

Continuous automatic analytical methods have also been described, including a determination by UV-abosrption.

A rapid method consists of bromination into tribromophenol and subsequent determination of the turbidity. Bromination with subsequent titration of the bromine consumption according to Koppeschaar is customary primarily for higher phenol concentrations; the brominated phenols can also be determined by IR-spectrophotometry as well as gas chromatography.

Because of the low stability of phenols in the presence of oxygen, water samples should be analyzed soon after they are obtained. For preservation for a few days, they are treated with NaOH to pH = 12 and stored in completely filled bottles with a glass stopper according to DEV. In contrast, ASTM recommends acidification with $H_3PO_4$ to pH = 4 and addition of 1 g $CuSO_4$· $5H_2O$ per liter.

## 9.2   Determination of Total Phenols by Bromination According to DEV (Phenol Content Higher Than 100 mg/l)

Of the neutral or weakly acid (pH = 5) water sample 500 ml are extracted three times with 100 ml portions of benzeen-quinoline mixture (4 + 1 by vol.). The organic phase is extracted twice with 100 ml portions of NaOH (d = 1.15) for 5 min each time. The NaOH extract is clarified by brief evaporation, extracted with 100 ml carbon tetrachloride and the tetrachloride extract is discharged. The phenolate solution is brought to 500 ml with water in a graduated flask.

Of this solution 50 ml are placed into a ground-joint Erlenmeyer flask with an attached dropping funnel in which they are treated with exactly 15 ml 0.1 N potassium-bromide-bromate solution and treated through the dropping funnel with 10 ml sulfuric acid (d = 1.22). After standing for 1 hour, 10 ml 10% potassium iodide solution are added through the dropping funnel. After standing for 10 min the liberated iodine is titrated with 0.1 N $Na_2S_2O_3$ solution. A blank experiment with 500 ml distilled water is carried out by the same method. If the difference between blank consumption and analysis exceed 10 ml $Na_2S_2O_3$ solution, 50 ml of the phenolate solution are diluted to 10 times the volume before the bromination and 50 ml are removed. One ml 0.1 N $Na_2S_2O_3$ solution corresponds to 1.7 mg total phenols (average molecular weight 102).

Reagents: NaOH, d = 1.15: 160 g NaOH are dissolved in 1 l of water.

$H_2SO_4$,d = 1.22: 230 ml concentrated $H_2SO_4$ are added to 1 l of water.

0.1 N potassium bromide-bromate solution: 2.784 dried $KBrO_3$, analytical grade, + 10 g KBr per liter.

Puschel and Grubitsch[253] use butylacetates as the extractant which is also applied in industry in the extraction of phenols from gas water (Phenolsolvan process). After adding $CuSO_4$ solution, they extract 50 ml of the weakly acid waste water sample twice with 40 ml butylacetate, extract the butylacetate extracts three times with 30 ml portions of 10% NaOH, and titrate an aliquot of the aqueous-alkaline phenolate solution as in the above with bromide-bromates.

For the bromination process in homologous phenols, see Erichsen and Rudolphi[77a] as well as Thielemann[307].

*Volatile phenols according to DEV (phenol content more than 100 mg/l).* Of the water sample 100 ml are treated with $CuSO_4$ solution in order to bind $H_2S$ in distillation equipment with a descending condenser until the supernatant liquid over the precipitate has turned blue. Subsequently it is treated with dilute $H_2SO_4$ (d = 1.15) until any precipitated blue copper hydroxide has gone into solution. Distillation is carried out into a water receiver up to a 20 ml residue; the latter is diluted with 50 ml water and again concentrated to 20 ml. The distillates are brought to 500 ml in a graduated flask. Of this amount, 100 ml are brominated as described above with 0.1 N $KBrO_3$-KBr solution and titrated with 0.1 N $Na_2S_2O_3$ solution.

$CuSO_4$ solution: 110 g $CuSO_4 \cdot 5 H_2O$ per liter of water.

$H_2SO_4$ (d = 1.15): 150 ml concentrated $H_2SO_4$ are poured into 1 l of water.

## 9.3 Colorimetric Methods

### 9.3.1 Colorimetric Determination with p-Nitraniline (DEV) for Phenol Concentrations of more than 0.1 mg/l

In the calibration of the technique with phenol the homologous phenols are determined to varying degrees: 100% phenol, 147% o-cresol, 120% m-cresol, 21% p-cresol, 16% o-xylenol, 52% m-xylenol, 92% p-xylenol.

In distillation equipment 200 ml of the water sample are treated with 1 ml $CuSO_4$ solution and 1 ml $CoSO_4$ solution (to bind the cyanides) and acidified with 10 ml phosphoric acid (d = 1.7). A receiver containing 10 ml N $Na_2CO_3$ solution is used to distill up to a residue of 20 ml and the distillate is diluted to 250 ml.

In a 100 ml graduated flask 20 ml p-nitraniline solution are treated with a few drops of saturated $NaNO_2$ solution until decolorization is obtained. Up to 50 ml of the distillate are mixed with 30 ml N $Na_2CO_3$ solution, added to the diazo-solution and brought to 100 ml. After 20 min, the solution is read in a photometer at 530 nm against a blank solution prepared by the same method. A calibration curve is constructed with solutions of 0.1 — 10 mg phenol/l.

$CuSO_4$ solution: 10 g $CuSO_4 \cdot 5\ H_2O$ and 100 ml water.

$CoSO_4$ solution: 10 g $CoSO_4 \cdot H_2O$ and 100 ml water.

p-Nitraniline solution: 1.38 g p-nitraniline are dissolved in 310 ml N HCl and brought to 2 l with water.

*Extraction of dye with butanol (for phenol concentrations of less than 0.2 mg/l).* The distillate obtained as above from 200 ml of water sample (receiver containing 30 ml N $Na_2CO_3$ solution) is treated with 20 ml diazotized p-nitraniline solution in a 500 ml separatory funnel. After 20 min the formed dye is extracted with 50 ml n-butanol and butanol solution is read in the photometer at 530 nm. The calibration curve is constructed with solutions of 0.1 — 0.4 mg phenol/l.

### 9.3.2   *Phenol Determination with Aminoantipyrine*

This method is described in several editions of APHA Standard Methods as well as ASTM D 1783; it can also be found in Water Analysis, a brochure of Merck Company. The sensitivity of the method extends into the ppb range. As mentioned above, this method does not respond equally to all phenol homologs, as is true for other colorimetric methods. Some p-substituents, for example, p-cresol, are not indicated by it.

Sample preparation: for preservation the water sample is acidified with $H_3PO_4$ (pH = 4) and treated with 1 g $CuSO_4 \cdot 5\ H_2O$ per liter in order to destroy bacteria and precipitate $H_2S$. Oxidizing agents (for example, free chlorine) are eliminated by adding a ferricyanide.

Distillation: From an acidified water sample (pH = 4) of 500 ml, 450 ml are distilled away. The residue is diluted with 50 ml water; another 50 ml are distilled away and the distillates aer combined.

*Direct Photometry (Phenol concentrations of 0.1 — 5 mg/l).* Of the distillates 100 ml are treated with 2 ml of 5% $NH_4Cl$ solution, adjusted to pH = 10 ± 0.2 with concentrated ammonia, and treated with 2 ml 2% freshly prepared aqueous aminoantipyrine solution and 2 ml 8% $K_3Fe(CN)_6$ solution. After 15 min, the solution is read in the photometer at 510 nm against a blank solution. The calibration curve is constructed with phenol solutions containing 0 — 0.5 mg phenol per 100 ml.

*Photometry after dye extraction $CHCl_3$ (phenol concentration of 0.05 — 0.1 mg/l).* Five hundred ml of distillate ($<$0.1 mg/l) or an aliquot diluted to 500 ml are treated with 10 ml of 5% $NH_4Cl$ solution, adjusted ot pH $= 10 \pm 0.2$ with concentrated $NH_3$ and treated with 3 ml 2% freshly prepared aminoantipyrine solution and 3 ml 8% $K_3Fe$ $(CN)_6$ solution in a 1 l separatory funnel; after 3 min the solution is extracted with 50 ml chloroform (for a 100 mm cuvette). The chloroform solution is filtered over anhydrous $Na_2SO_4$ and read in the photometer at 460 nm. The calibration curve is constructed with 5—50 $\mu$g phenol/500 ml.

*Determination without distillation.* A water sample of 100 — 1000 ml is acidified with 10 ml concentrated HCl and extracted three times with 50 ml portions of petroleum ether. The phenols from the combined petroleum ether extracts are extracted first with 10 ml and then twice with 5 ml portions of 0.5 N $NH_3$. The pH is adjusted to 7.9 with phosphate buffer (pH $= 6.8$), 0.50 ml of the above aminoantipyrine solution and 0.50 ml $K_3Fe(CN)_6$ solution are added, and after 15 min a photometer reading is taken at 500 nm. The calibration curve is constructed in the same manner.

With regard to the pH-dependence of the phenol reaction with antipyrine see also Faust and Mikulewicz[84].

### 9.3.3    Colorimetric Determination with Gibbs Reagent

The determination with Gibbs reagent (2,6-dibromoquinone-4-chlorimide, obtainable from Merck) is described in the tenth edition (1955) of the APHA Standard Methods, and elsewhere. See also Naucke[236a].

Of the water sample, containing not more than 10.1 mg phenol/l, 300 ml are adjusted to pH $= 9.4 \pm 0.2$ with 15 ml borate buffer. Five ml Gibbs reagent are added and the solution is allowed to stand for 6-24 h. The dye is extracted with 75 ml n-butanol, followed by photometry at 670 nm.

Borate buffer: 3.1 g $H_3BO_3$ and 3.5 g KCl are treated with 32 ml N NaOH and diluted to 1 l. A dilution from 5 ml to 100 ml should show a pH value of 9.4 $\pm$ 0.2.

Gibbs reagent: 0.2 g 2,6-dibromoquinone-4-chlorimide (Merck) is dissolved in 50 ml 95% alcohol. This solution is stable for about one week. Distilled water is used to dilute 4.5 ml to 100 ml. This dilution is stable for only about 30 min.

A procedure on a microscale with 1 ml water sample has been described by Gorbach, Koch and Dedic[113].

Dacre[62] has reported on the characteristics of 50 homologous and substituted phenols in the presence of Gibbs reagent.

### 9.3.4    Phenol Determination with Folin-Denis Reagent

According to the first edition of DEV, phenol concentrations of 0.1 — 2 mg/l are analyzed by treating 100 ml distillate with 10 ml cold-saturated soda solution and 2 ml phenol reagent. After standing for 1 h, the blue color is compared with simultaneously prepared phenol test solutions but can also be analyzed by colorimetry with an S 66 filter.

Phenol reagent: 100 g sodium tungstate are dissolved in 750 ml water and a solution 35 g $MoO_3$ in 50 ml 85% phosphoric acid is added. The solution is boiled for 2 h with refluxing, a solution of 50 mg uranylacetate in 50 ml water is added and the solution is brought to 1 l. After standing for 8 days, the solution is ready for use and is stable in a brown bottle.

The reaction is not specific for phenols; it also occurs in sulfite waste effluents, for example.

### 9.3.5    Phenol Determination by Nitrogenation

A simple total phenol determination by nitrogenation has been described by Morkowski[232]. A water sample of 100 ml is boiled with refluxing for 10 min with 10 ml 65% nitric acid (d = 1.4). After cooling 20 ml 25% ammonia solution are added and the yellow color is analyzed by photometry at 420 nm. Limit of detection: 0.1 mg/l.

### 9.4    Phenol Determination by Bromination and IR-Spectroscopy  (Simard et al.[290])

A water sample of 1 l is treated with 100 g KBr, 25 ml $KBrO_3$ solution (12 g/l) and 80 ml HCl (1 + 3). The mixture is shaken for 5 min, the bromine excess is eliminated by addition of 30 ml 10% $Na_2S_2O_3$ solution, the formed tribromophenol is taken up with 20-100 ml $CCl_4$ by shaking for 15 min, the quantity being a function of the expected phenol concentration, i.e. 0.01 — 0.05 ppm per liter of water sample. For elimination of acids, the $CCl_4$ extract is shaken for 5 min with 125 ml of 2% $NaHCO_3$ solution. The $CCl_4$ solution is filtered through paper into a 50 mm glass cuvette with a quartz window and the absorption is read at 2.84 $\mu$m in an IR-spectrophotometer with a lithium fluoride prism. The limit of detection is 0.01 ppm phenol.

Mohler and Jacob[224] have compared the advantages and disadvantages of five methods of phenol determination (IR, UV, Gibbs, 4-aminoantipyrine, nitrosophenol).

### 9.5 Phenol Determination by Measurement of the UV-Difference in Acid and Alkaline Solution

Phenol and its homologs exhibit a strong increase of light absorption in alkaline solution compared to their acid solution. This serves as the basis for the analytical method of Schmauch and Grubb[274] (see also Wexler[332] as well as Martin et al[217]).

Procedure: a water sample of 900 ml is adjusted to pH = 12 with KOH. Ten g NaCl are added, followed by extraction with 10 ml $CCl_4$ for 30 min. If necessary, the aqueous — alkaline layer is filtered, adjusted to pH = 5 with HCl and shaken for 30 min with 10 ml tributylphosphate. In a 5 ml graduated flask, 4 ml of the organic phase are alkalinized with 1 ml of 0.1 N tetrabutylammonium hydroxide solution in methanol. The absorbance A is read at 301 nm in a 1 cm quartz cuvette. A second 4 ml portion of the tributylphosphate solution, brought to 5 ml with methanol in the absence of alkali, serves as a reference solution. If the absorbance exceeds a value of 0.8, both solutions are diluted in the same proportion, *i.e.,* the alkaline solution with a mixture of 4 vol. tributylphosphate and 1 vol. of the 0.1 N quaternary base in methanol, and the acid solution with tributylphosphate/methanol, 4 + 1.

An absorbance correction $A_c$ for cuvette and reagents is obtained by measuring the absorbance of the alkaline 4 + 1 mixture against the tributylphosphate + methanol mixture.

The phenol concentration C (mg/l) is obtained from the formula:

$$C = 12500 \, (A - A_c) \cdot F/V \cdot a \qquad (1)$$

where S is the addition factor and a is a calibration constant obtained by reading the absorbance of weighted phenol samples in tributyl phosphate according to the formula a $= 1.25$ $(a - a_c)/P$ (a = absorbance of standard solution, $a_c$ = absorbance correction, P = phenol concentration of the standard solution in g/l).

An absorbance of 0.015 corresponds to a phenol concentration of 10 ppb of which 65% can be detected.

## 9.6 Thin-Layer Chromatography (TLC)

*TLC-separation of phenols according to Dietz and Koppe*[63]. After extraction and concentration of the ether solution according to p. 115, the phenols, concentrated to about 20 $\mu$l, are taken up in small amounts of dioxane and brought to 100 $\mu$l in a graduated microflask with a microsyringe. Fifty $\mu$l of this solution are applied on the TLC plate and the phenols are released at the start spot by addition dioxane-glacial acetic acid (4 + 1).

A satisfactory separation can be realized by the two-dimensional technique on silica gel plates. Benzene is the solvent in the first dimension; for the second dimension, a 4 cm strip, serving for development in the first dimension, is covered with plastic film, while the rest of the plate is sprayed with 2 N $Na_2CO_3$ solution and developed with diisopropylether.

Diazotized sulfanilic acid serves as the detection agent.

Heier[128] uses $Na_2CO_3$-alkalinized silica gel G, a solvent system of toluene/ethanol/acetone (120: 12 : 1), and a spray detector of p-nitraniline in N HCl, decolorized with $NaNO_2$. The spots can be analyzed with suitable recording instruments.

Low phenol concentrations in ground and surface water can be enriched first on activated carbon according to Thielemann[309]. They are then extracted with ether, concentrated, taken up in water and coupled with a pinch of Fast Red Salt AL; after 30 min, they are acidified with 2 N HCl and the dye is extracted with chloroform, followed by concentration. The following are used for the TLC separation: $K_2CO_3$-alkalinized silica gel G plates (activated for 2 h at 100°); solvent system: either dichloromethane/ethylacetate/diethylamine (92:5:3) or chloroform/ethylacetate/methanol (93:5:2) or benzene.

Thielemann[308] separates chlorophenols by TLC on silica gel G plates activated for 2 h at 100°. Benzene serves as the solvent and diazotized sulfanic acid as the detector.

*TLC determination of chlorophenol.* Zigler and Phililps[346] determined chlorophenols by two-directional TLC. In the first direction benzene eliminates extraneous materials, while the chlorophenols are separated in the second direction. Limit of detection: 0.1 $\mu$/l.

Extraction: a water sample of 1 l is acidified to pH = 1.5 with $H_3PO_4$ and extracted four times with 100 ml portions of petroleum ether. The petroleum ether solution is concentrated to 5 — 10 ml, dried with anhydrous $Na_2SO_4$, centrifuged and concentrated to 0.1 ml.

TLC: $Al_2O_3$ G, dried at 120° for 1 h.

The petroleum ether extract is applied in one corner (2 x 2 cm distance from edge), and is purified and dried with benzene in the ascending direction for a distance of 9 — 10 cm. The chlorophenols do not move. Solvent for the other direction: 6 ml N NaOH and 94 ml acetone. Development is continued up to 2 cm from the top edge. Chlorine is detected by spraying with $AgNO_3$ solution (0.5 g $AgNO_3$ dissolved in 5 ml water, mixed with 100 ml 2-phenoxyethanol and brought to 1 l with acetone, treated with 3 drops 30% $H_2O_2$ and stored in a brown bottle). The plate is dried for 1-2 min at 80° and exposed to UV-irradiation for 15 min. Chlorine-containing spots turn brown.

*Identification of phenols:* the plate is sprayed with 2% solution of 4-aminoantipyrine in acetone, dried in a gentle warm air stream, sprayed with aqueous N NaOH, dried again with hot air, and finally sprayed with 8% $K_3Fe(CN)_6$. Phenols form pink-red spots on a yellow background.

Rf-values: m-chlorophenol 0.94; 2,4-dichlorophenol 0.71; 2,4,5-trichlorophenol 0.62; 2,4,6-trichlorophenol 0.42; pentachlorophenol 0.09.

Seeboth[282a] separated phenols by TLC on silica gel G. Four solvents are described: $CHCl_3$-glacial acetic acid 5:1; $CHCl_3$-acetone-glacial acetic acid 10:2:1; benzene-glacial acetic acid 5:1; petroleum ether (b.p. 60-80°)-$CCl_4$-glacial acetic acid 4:6:1. The detection agent is a spray of p-nitrobenzene-diazonium fluoroborate. The development time with solvent system 3 amounts to 50 min (15 cm).

Seeboth and Gorsch[282b] have described the quantitative analysis.

Aly[8] reported on the TLC determination of phenols after their reaction with diazotized p-nitrophenol and aminoantipyrine.

## 9.7   Gas Chromatography (GC)

*GC phenol determination by direct injection according to ASTM D 2580 (see also Baker* [15a]*).* The specification of ASTM D 2580 contains a detailed description of a gas-chromatographic phenol determination (phenol, cresols, mono- and dichlorophenols) by direct injection of the water sample without previous isolation and enrichment procedures; the method includes precise information on the equipment and procedure of the analysis, including calibration, retention times and relative peak areas of the individual phenols, as well as the obtainable reproducibility. Limit of detection: 1 mg/l.

If the water sample is to be injected directly into the gas chromatograph, it may not contain suspended particles which would plug the injection needle. Non-phenolic components with a similar retention behavior as phenols are an interference; in these cases pretreatment, for example, according to p. 115, is necessary. Residual peaks (ghosts) from previous determinations are eliminated best by injecting 3 $\mu$l water with adjustment of the instrument to its full detection sensitivity.

GC equipment: temperature adjustment to 210° $\pm$ 0.2°, FID, 10 $\mu$l syringe, recorder with 1 mV deflection, response time 1 sec., paper feed 30 cm/h.

Columns: 3 mm inside diameter, stainless steel. Three column packings are described:

1. 60-80 mesh Chromosorb W, washed with acid and treated with HMDS, with 20% Carbowax 20 M-TPA (terephthalic acid), 3 m length.

2. 70-80 mesh Chomosorb W, washed with acid, wtih 5% free fatty acid phase (Varian Aerograph,) column length 1.5 m.

3. 60-80 mesh Chromosorb T (Teflon-preparation of m.p. 327°) with 10% free fatty acid phase; column length 3 m.

Column conditioning for 24 h at 250°. Sample volume 1 $\mu$l.

Operating conditions

| Column-number | 1 | 2 | 3 |
|---|---|---|---|
| Carrier gas | He | He | $N_2$ |
| Carrier gas, flow rate, ml/min | 25 | 35 | 60 |
| Sample inlet temperature | 250° | 205° | 250° |
| Column temperature | 210° | 147° | 188° |
| Hydrogen for FTD ml/min | 25 | 25 | 30 |

Relative retention times

| Phenol | Boiling pt. | Column 1 | Column 3 |
|---|---|---|---|
| Phenol | 182° | 1.0 | 1.0 |
| o-Cresol | 192° | 1.0 | 1.0 |
| m-Cresol | 203° | 1.3 | 1.3 |
| p-Cresol | 202° | 1.3 | 1.3 |
| o-Chlorophenol | 176° | 0.8 | 0.6 |
| m-Chlorophenol | 214° | 3.6 | 3.6 |
| p-Chlorophenol | 217° | 3.6 | 3.6 |
| 2,3-Dichlorophenol | | 1.8 | 1.9 |
| 2,4-Dichlorophenol | 210° | 1.8 | 1.9 |
| 2,5-Dichlorophenol | 210° | 1.8 | 1.9 |
| 2,6-Dichlorophenol | 220° | 1.6 | 1.5 |

Comparison of results with gas chromatography and with aminoantipyrine

| | Concentration (mg/l) | | |
| | Sample weight | GC | Aminoantipyrine |
|---|---|---|---|
| Phenol | 1.06 | 0.97 | 0.97 |
| o-Cresol | 1.04 | 1.03 | 0.64 |
| m-Cresol | 1.02 | 1.03 | 0.38 |
| p-Cresol | 1.00 | 1.00 | 0.00 |
| Mixture | 4.12 | 3.9-4.1 | 2.40 |

*Enrichment and GC according to Scholz*[277]. Enrichment: a water sample of 2.5 l is alkalinized with 2.5 g KOH (pH = 12) and concentrated in vacuum to 30 — 50 ml (temperature not more than 20°). The sample is acidified to pH = 1-3 with HCl and, after adding 7.5 g NaCl, is extracted with 30-50 ml ethylacetate. The extract is dried with anhydrous $Na_2SO_4$, treated with 250 μl benzyl benzoate and the ethylaetate is evaporated in vacuum at 15 torr.

Gas chromatography: Sample volume: 5 μl concentrate; Perkin Elmer Fractometer 116E; furnace temperature 160°; 1 m column; stationary phase: Apiezon grease L on Celite; carrier gas; helium 50 ml/min, FID. The phenols emerge in 15 min and the benzyl benzoate only after 12 h. Thus, 4-6 samples can be analyzed before the ester peak is eluted. The ester is then outgassed at higher temperature.

*GC-Determination according to Nauke and Tackmann*[237]. For the determination of higher phenol concentrations (>25 mg/l) Nauke and Tackmann extract a water sample of 100 ml at pH = 3-5 with exactly 10 ml isobutylacetate, dry the ester extract on anhydrous $Na_2SO_4$ and inject 20 $\mu$l of the solution into a **Perkin Elmer** Fractometer 116 with a thermistor detector. The carrier gas is helium 55 ml/min; 4 m column; 20% Apiezon grease L on Celite; furnace temperature 180°. Relative retention time (by extract, phenol = 1, after 10 min): o-cresol 1.53; m + p-cresol 1.63; dimethylphenols 2.2-3,1; trimethylphenols 3.8-4.

Lower phenol concentrations can be extracted with ether. The ether extract is concentrated to 0.5 ml.

Sementschenko and Kaplin[284] have described the GC analysis of phenols after their methylation or acetylation.

*GC phenol determination in the CCE extract.* Eichelberger, Dressman and Longbottom[74] determined phenols by gas chromatography in the activated carbon $CHCl_3$ extract. The carbonchloroform extract CCE) is evaporated to 30 ml instead of to dryness. This quantity is extracted three times with 50 ml portions of NaOH (pH = 13). The alkaline extracts are acidified with concentrated HCl to pH = 2 and extracted three times with 15 ml portions of ethylether. The ether solution is dried with anhydrous $Na_2SO_4$ and allowed to flow through a small column with Florisil; the column is rinsed with 200 ml ether and the sample is concentrated to 50 ml on a water bath of 50°. Of this sample, 5 $\mu$l are injected into the gas chromatograph.

GC-conditions: Perkin Elmer Model 800 with FID; 3 m column

Retention times relative to phenol = 1.

| Column temperature | | | |
|---|---|---|---|
| o-Chlorophenol | 0.63 | 2,3-Dimethylphenol | 1.54 |
| o-Nitrophenol | 0.72 | 2,4-Dichlorophenol | 1.70 |
| 2,6-Dimethylphenol | 0.78 | 3,5-Dimethylphenol | 1.71 |
| 2,6-Ditert.-Butyl-p-Cresol | 0.83 | 2,5-Dichlorophenol | 1.76 |
| o-Cresol | 0.98 | 2,3-Dichlorophenol | 1.78 |
| Phenol | 1.00 | 2,3,5,6-Tetramethylphenol | 1.83 |
| o-Aminophenol | 1.21 | 2,4,5-Trimethylphenol | 1.88 |
| 2,5-Dimethylphenol | 1.24 | 2,3,5-Trimethylphenol | 1.96 |
| 2,4-Dimethylphenol | 1.26 | 3,4-Dimethylphenol | 1.99 |
| p-Cresol | 1.27 | p-Tert.-Butylphenol | 2.43 |
| m-Cresol | 1.30 | 2,4,6-Trichlorophenol | 2.88 |
| 4-Chloro-2-Nitrophenol | 1.32 | p-Methoxyphenol | 3.32 |
| 2,6-Dichlorophenol | 1.40 | p-Chlorophenol | 3.65 |

| Column temperature 240°. | | | |
|---|---|---|---|
| m-Chlorophenol | 3.12 | 1-Naphthol | 10.90 |
| 2,4,5-Trichlorophenol | 4.09 | 2-Naphthol | 12.51 |
| 3,4-Dichlorophenol | 9.64 | | |

of 3 mm I.D., packed with HMDS-treated Chromosorb W (60 — 80 mesh), with 10% Carbowax 20M + terephthalic acid; carrier gas $N_2$, 50 ml/min; temperature: 260° in the injector and 210° in the detector.

*Gaschromatography of phenols after their conversion into halogenated derivatives.* The sensitivity of a gas-chromatographic determination can be enhanced considerably by converting the phenols into halogenated derivatives which are then analyzed with an electron capture detector.

The simplest method is bromination with bromide-bromate as described in 9.4. According to Umbreit, Nygren and Testa[316] the bromophenols are extracted with issooctane and subjected to gas chromatography and analyzed with an electron capture detector. Samples with a phenol concentration of 5-15 ppb could be analyzed with a standard deviation of ± 1 ppb.

Kawahara[159] used a similar method to convert the weakly acid phenol-containing fraction of the carbon-chloroform extract with α-bromopentafluorotoluene into the corresponding pentafluoro-benzylether which is then determined by gas chromatography with an electron capture detector (ECD).

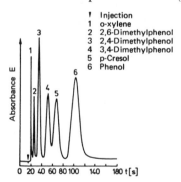

Fig. 9.8—Liquid-liquid chromatogram of phenols

According to Argauer[11] acetylation with chloroacetic anhydride also yields a derivative suitable for ECD identification.

## 9.8 Phenol Determination by High-Efficiency Liquid-Liquid Chromatography

The separation of a mixture of phenol homologs by liquid chromatography (Siemens equipment) is described in Fig. 9.8.

| Carrier flow rate: | 5 ml/min | Adjusted wave length on the detector: | $\lambda = 270$ nm |
|---|---|---|---|
| Separating column: | | Thermostat-temperature: | 20° C |
| Lengths | 100 mm | Stationary phase: | 5% Fraktonitril III and Chomosorb G |
| I.D. | 3 mm | Mobile phase: | Isooctane |

# 10. DETERGENTS (SURFACTANTS)

## 10.1 Summary

Detergents or surfactants are agents with cleaning power which reduce the surface tension of water. Their presence is evident externally by the formation of foams when the water is moved or shaken. The most important applications of these compounds are as washing, cleaning and rinsing agents.

The chemical structure of the compounds consists of a combination of hydrophobic and hydrophylic groups of atoms. A distinction is made between anionic, cationic and nonionic surfactants which are treated separately in analyses. The most important surfactants contained in detergent powders for home use are anionic. For special purposes, especially as disinfectants, cationic agents are sometimes used. Nonionic surfactants also occur in nature, for example, the saponins which have been used occasionally as washing agents. Synthetic neutral compounds (nonionics) are finding increasing use in detergents because of several advantages.

The diverse use of synthetic surfactants for washing of textiles, dishwashing and cleaning of motor vehicles as well as in cosmetic products has led to their increasing occurrence in residential and trade waste effluents. In contrast to soaps, they are stable to the water minerals (calcium and magnesium salts) and therefore do not enter the sludge phase of waste water; instead, they usually remain in solution while maintaining other water-insoluble pollutants in an emulsion-like distribution.

Surfactants have various harmful effects in waste and surface water. In the past, foaming often assumed dramatic proportions, particularly below dams in the form of foam layers with a height of several meters which inhibited ship travel. Because of their enrichment on the surface of the water they inhibited mass transfer; furthermore, they usually have a toxic effect on marine life and consequently reduce the self-cleaning power of water.

Many commercial products, especially tetrapropylenebenzenesulfonate, of which the highly branched aliphatic component can be manufactured from propylene in a relatively simple and inexpensive process, had the disadvantage of being stable in biological treatment processes and therefore becoming enriched in rivers in a very unfavorable manner. A law of 1962 prohibited all detergents with a biodegradability of less than 80% as determined in a precisely worded test method. The detergent industry quickly succeeded in the development of equally efficient, although somewhat more expensive biodegradable (biologically "soft") detergents. Thus, testing for biodegradability according to legally established standards has become an additional task of the water analyst.

## 10.2   Anionic Detergents

### 10.2.1   *Longwell and Maniece Methylene Blue Method*[204]

Most of the procedures based on the methylene blue method, for example, the German Standard Methods (H 23) are based on a study of Longwell and Maniece[204]. By a choice of the most favorable pH of 10 and a washing process with acid methylene blue solution, these authors were able to prevent a number of interferences from inorganic ions and protein compounds.

*Procedure:* A water sample of 100 ml containing 20-250 $\mu$g anionic substance is shaken in a 250 ml separatory funnel with 10 ml phosphate buffer (pH = 10), 5 ml *neutral* methylene blue solution and 15 ml chloroform for 1 min. The separated chloroform layer is shaken with a mixture of 100 ml distilled water and 5 ml *acid* methylene blue solution is a second separatory funnel, is allowed to settle and is filtered through a chloroform-moistened cotton filter into a 50 ml graduated flask. Extraction in both separatory funnels is repeated two more times, each time with 10 ml chloroform; the chloroform extracts are brought to 50 ml and are read in a spectrophotometer at 650 nm or a suitable filter photometer (for example, Elko II with interference filter J 66). A blank value is obtained by the same method.

Evalution is performed with a calibration curve with 25, 50 100, 150, 200 and 250 $\mu$g standard tetrapropylenebenzenesulfonate per 100 ml = 10, 20, 40, 60, 80 and 100 ml calibration solution diluted to 100 ml.

*Reagents:* Neutral methylene blue solution: 0.35 g methylene blue (DAB 6) is dissolved in 1 l of water and, after standing for 24 hr, is ready for use. The absorbance of the chloroform phase from the blank sample prepared with it should not exceed a value of 0.015 per cm of optical path.

Acid methylene blue solution: 0.35 g methylene blue is dissolved in about 500 ml water, treated with 6.5 ml concentrated sulfuric acid and brought to 1 l with water. This solution should also be used only after standing for 24 hr, the chloroform phase of the blank sample should have an absorbance of less than 0.015 per cm of optical path.

Phosphate buffer solution, pH = 10: 12.52 g dibasic sodium phosphate $Na_2HPO_4 \cdot 2 H_2O$ are dissolved in 500 ml water. About 3 ml of 0.5 N NaOH serve to adjust to pH = 10, followed by dilution with water to 1 l.

Standard solution: 20 ml of an exactly 5% tetrapropylenebenzenesulfonate solution (TBS)* of Merck Company are diluted to 1 l with water and 50 ml of this solution is diluted again to 1 l. Of this stable stock solution 50 ml are again diluted to 1 l. This solution contains 2.5 $\mu$g TBS/ml (ABS/ml).

If the water sample contains sulfide, the addition of the alkaline buffer solution is followed by 2 ml 20% hydrogen peroxide, and the solution is allowed to stand for 5 min.

To remove residues of surface-active cleaning agents, the glassware should be rinsed with a mixture of 9 parts methanol or ethanol and 1 part concentrated HCl.

According to Longwell and Maniece, less than 0.01 ppm is found as a blank value in clean water while 87-97% of the detergents added to residential waste water is recovered.

Abbott[1] recommends that the blank value be reduced by extracting the neutral, buffer-treated methylene blue reagent solution with chloroform immediately before addition to the water sample. In place of the phosphate buffer, he uses a borate buffer of equal volumes of 0.05 M $Na_2B_4O_7$ and 0.10 N NaOH. This measure improves the reproducibility of the results to ± 0.002 ppm and 98-100% of added test substances (TBS as well as dioctylsulfosuccinate) are found. He recommends the use of ethanol-free chloroform.

In order to simplify the method without a loss of accuracy, Slack[290a] recommends that the extraction be performed only once with exactly 50 ml chloroform. For further simplifications see Panowitz and Renn[245] as well as Brink[39].

---

*In U.S., usually called ABS (alkylbenzenesufonate)

Södergren[292] has described the automation of the procedure by means of the auto-analyzer.

Shaking equipment for the simultaneous treatment of 16 separatory funnels has been described by Niemitz and Fuss[239].

### 10.2.2    Use of Other Dyes

*Methyl green.* According to Moore and Kolbeson[229] methyl green is less subject to interference than methylene blue in a detergent analysis.

Procedure: A water sample of 20 ml in a separatory funnel is treated with 10 ml buffer solution (pH = 2.5) and 2 ml methyl green solution; it is shaken with exactly 40 ml benzene for 1 min (250 impulses). The sample is allowed to settle, the aqueous layer is siphoned away, the benzene layer is washed with a mixture of 15 ml water and 5 ml buffer solution, is allowed to stand for 20-30 min up to clarification, pipetted into a cuvette without filtration and absorbance read spectrophotometrically at 615 nm.

*Reagents:* Methyl green solution: 0.5 g of dye is dissolved in 100 ml water. Abbott recommends that the reagent solution be extracted with chloroform until the latter takes up no further color.

Buffer, pH = 2.5: 7.5 g glycocoll and 5.8 g NaCl are dissolved in water to 1 l and adjusted to pH 2.5 with 0.1 N HCl.

Standard solution: 5-100 $\mu$g Na laurylsulfate per 20 ml water, 50 $\mu$g in 20 ml water; 50 $\mu$g in 20 ml in a 1 cm cuvette, read at 615 nm, produce an absorbance of 0.20; standard deviation $\pm$ 0.3 $\mu$g = $\pm$ 0.015 ppm.

*Azure A.* Azure A produces intense colors with detergents according to Steveninck and Riemsma[298] as well as according to Tonkelaar and Bergshoeff[313]; it is therefore particularly suited for low concentrations.

Of the water sample, 50 ml (1-30 $\mu$g anionic material) are vigorously shaken with 5 ml 0.1 N $H_2SO_4$, 1 ml Azuse A solution and exactly 10 ml chloroform for 2 min. The chloroform layer is filtered through glass wool into a cuvette and read in the spectrophotometer at 623 nm against chloroform.

Azure A solution: 40 mg Azure A (Allied Chemical Corporation) are dissolved in 5 ml 0.1 N $H_2SO_4$ and diluted to 100 ml. A surfactant concentration of 5 $\mu$g in 50 ml water with Azure A produces an absorbance of 0.1 while a value of only 0.03 is obtained with methylene blue.

The use of ferroin intsead of methylene blue has been described by Taylor and Fryer[305].

With regard to the separation of anionic detergents by liquid chromatography on a Sephadex G 10 column, see Mutter[236].

See Lovell and Sebba[205] on the use of flotation methods for surfactant analysis.

Le Bihan and Courtot-Coupez[182] have described the determination of anionic surfactants with Cu(II) phenanthroline, extraction of the complex with methylisobutylketone and determination of copper by atomic absorption flame photometry.

### *10.2.3    ASTM Method D 2330*

**(See also Fairing and Short[79] as well as Webster and Halliday[324]**

This method can be used if only C-sulfonates, particularly alkylarylsulfonate, are to be determined. Detergents which hydrolyze with strong acids, for example, alkylalcohol sulfuric acid esters, are decomposed by boiling with HCl.

A water sample containing not more than about 100 $\mu$g surfactant is boiled with 30 ml concentrated hydrochloric acid for about 1 hr and in the process is evaporated almost to dryness. The sample is diluted to 25 ml with water, treated with a few drops of phenolphthalein, made alkaline wtih NaOH, boiled for 2 min, rinsed into a separatory funnel, diluted to 100 ml, decolorized with sulfuric acid, treated with 10 ml buffer solution, pH = 7.5, extracted 4 times, each time with 25 ml of a mixture of 100 ml chloroform and 4 drops methylheptylamine; the chloroform extracts are evaporated in the presence of 5 ml 10% NaOH solution, the aqueous-alkaline residue is diluted to 100 ml and finally boiled for 15 min until the amine has completely evaporated. After addition of phenolphthalein, the residue is neutralized with dilute $H_2SO_4$ and the surfactant concentration is determined with methylene blue according to Longwell and Maniece.

Phosphate buffer, pH = 7.5: 10 g $KH_2PO_4$ are dissolved in 800 ml water, adjusted to pH = 7.5 $\pm$ 0.1 with NaOH solution and diluted to 1 l.

*IR-spectrophotometry of arylaklylsulfonates.* The APHA Standard Methods contain 2 procedures for the analysis of anionic detergents: a simplified methylene blue method and a somewhat more complicated IR-spectrophotometric method based on a publication of Sallee et al.[267]. Enrichment is performed with

100 g activated carbon, followed by drying at 105°, desorption by boiling with a mixture of 500 ml benzene, 420 ml methanol and 80 ml 0.5 N KOH. After separation of the neutral extracted materials wtih petroleum ether, the acid surfactants are extracted with chloroform-methylheptylamine, the chloroform is evaporated, the residue is taken up in $CS_2$ and read in the IR-spectrophotometer at between 9.0 and 10.5 μm. The peaks at 9.6 and 9.9 μm serve for the evaluation.

A somewhat simplified method which also permits a distinction of biodegradable (alkylarylsulfonate with a linear alkyl rest) and non-degradable (branched alkyl rest) surfactants, has been described by Ogden, Webster and Halliday[241a].

An absorption tube of 50 cm length and 2.5 cm I.D. is packed with 25 g 30-mesh activated carbon (Nuchar C 190). A water sample containing about 100 mg surfactant is loaded onto the column at a rate of 2 l/hr. Desorption is performed by pouring on a 2.5 l mixture of 2 l methanol, 500 ml chloroform and 25 ml ammonia water (d = 0.88) at room temperature at a rate of 25-30 ml/min. The solvent is evaporated, the residue is dissolved in 100 ml water, is boiled for 1 hr with 25 ml concentrated HCl, rinsed into a separatory funnel with 2 x 10 ml ethanol and the neutral components are extracted with 50 ml petroleum ether. The petroleum ether extract is extracted twice with 30 ml of 50 vol.% ethanol and is discarded. The aqueous-alcoholic solutions are treated with 1 ml n-heptylamine or 1-methylheptylamine, 70 ml water and sufficient ethanol so that the ethanol-water ratio is adjusted to 1:4. Subsequently, the material is extracted with 100 ml, and 4 times with 50 ml, petroleum ether (b.p. 40-60°). The petroleum ether is evaporated, the residue is dissolved in methanol and rinsed into a 50 ml graduated flask.

Of the methanol solution 5 and 25 ml are evaporated, respectively. The residue of the 5 ml sample is transferred into a 5 ml test tube and brought to the mark with $CS_2$. The total surfactant is obtained by IR-spectrophotometry in a 0.8 mm cuvette over the range of 9.0-10.5 μm with the reading being taken on the 9.9 μm band. The evaporated 25 ml aliquot is brought to the mark with $CCl_4$ in a 2 ml graduated flask and a reading is taken between 7.0 and 7.5 μm. The 7.31-μm band (see Fig. 10.2.3) corresponds to the linear (biodegradable) surfactants.

See also Mähler, Cripps and Greenberg[214] for the IR-spectrophotometry of degradable and non-degradable surfactants.

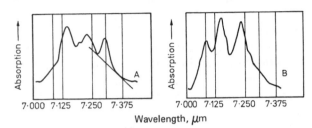

Wavelength, μm

Fig. 10.2.3.—IR-absorption spectra of l-methylheptylammoniumsalts of alkyl-benzenesulfonates: A = non-biodegradable surfactant, B = biodegradable surfactant (according to Ogden, Webster and Halliday).

## 10.3  Determination of Cationic Detergents with Bromphenol Blue (DEV, H 23)

Procedure: A water sample of 100 ml is shaken for 3 min in a separatory funnel with 10 ml citrate buffer solution, 5 ml 0.1 N HCl, 2 ml bromphenol blue solution and 50 ml chloroform. The chloroform layer is filtered over a small plug of cotton. The first 5 ml are discarded, while the balance is read spectrophotometrically at 416 nm. A calibration curve is constructed with suitably diluted standard solutions.

The method is suitable for concentrations of more than 0.05 mg/l cationic agents.

Citrate buffer solution: 21 g citric acid $(C_6H_8O_7 \cdot H_2O)$ are treated with 200 ml N NaOH and then diluted to 1 l with water. Of this solution 309 ml are diluted to 1 l with 0.1 N HCl.

Bromphenol blue solution: 0.150 g bromphenol blue are dissolved in 200 ml 0.01 N NaOH and treated with 42 ml 0.1 N HCl.

Standard solution: 1.00 g cetyltrimethylammonium bromide is dissolved with water to 1 l. Of this solution 50 ml are diluted to 1 l, and from this dilution, 50 ml are again brought to 1 l with water. One ml of this standard solution contains 2.5 μg cetyltrimethylammonium bromide.

For an analysis of the content of the preparation utilized, 2.00 g are dissolved in water and brought to 1 l. Of this solution 100 ml are precipitated with 25 ml 0.1 N potassium dichromate solution, allowed to stand for 2 hr, filtered through a coarse membrane filter and, after addition of 5 g potassium iodide and 10 ml sulfuric acid (3 vol. water + 1 vol. concentrated $H_2SO_4$) and standing for 10 min, the chromate excess in the clear filtrate is titrated with 0.1 N $Na_2S_2O_3$ and zinc iodide-starch solution.

The percentage G of cationic substance is obtained according to the formula:

$$G = \frac{(a-b) \cdot A \cdot 0.333}{c}$$

where a is the consumption of 0.1 N $Na_2S_2O_3$ solution in ml in the blank experiment and b is the corresponding consumption in the sample, A is the equivalent weight of cationic agent (364 in the case of cetyltrimethylammonium bromide), while c represents the sample weight in g.

Sheiham and Pinfold[285] report on the use of citric acid instead of bromphenol blue.

## 10.4    Nonionic Surfactants (Nonionics)

The nonionic surfactants also consist of a combination of water- and fat-soluble molecular fractions. The water-soluble fraction as a rule consists of ethylene oxide adducts [($\cdot O \cdot CH_2 \cdot CH_2 \cdot O$)-chains], while the fat-soluble or hydrophobic fraction is represented by arylalkyl rests with a longer aliphatic side chain. Nonylphenol with a chain of 10 mol ethylene oxide (for example, Marlophen 810) can serve as a model substance.

Methods of determination for substances of this group as a rule require the separation of interfering anionic surfactants. This separation can be performed with anion exchangers or the non-ionic surfactants are isolated by extraction or defoaming methods.

For the actual determination a number of inorganic complex formers is available, but often these are not strictly specific because of the inclusion of different types of materials. A survey of these reagents has been published by Heinerth[129] as well as Wickbold[334].

Cobalt rhodanide (Browns and Hayes[40], Crabb and Persinger[57], Greff, Setzkorn and Leslie[115], Huddleston and Allred[140], Sebban[282], Milwidsky[221]) as well as barium-tungsten iodide (Bürger[45], Wickbold[334]) are used most frequently. Various versions of determination have been described with both methods.

*Hey and Jenkins method*[136]. The authors isolate nonionic surfactants from the water sample by defoaming. Air is blown through the sample water; the air bubbles are conducted through a benzene layer which takes up the surfactants. The surfactant adds to cobalt rhodanide, the complex is decomposed and the rhodanide fraction is determined by spectrophotometry according to Bark and Higson.

Suitable equipment for the purpose consists of a glass tube (40 cm long, 2 cm inside diameter) with 2 marks which form a volume of 15 ml. The upper mark is 5 cm from the top rim which contains a spout. A sintered glass disk (porosity 2) is tightly fitted in at the lower edge and an air stream can be blown through it. The tube is charged with 100 ml water sample which has been alkalinized with 10 drops of N NaOH. The air stream is adjusted to about 35-40 ml/min. Benzene (15-20 ml) is layered onto the aqueous layer. After 3 hr of air blowing, sufficent water is added that the benzene layer becomes measurable between the marks and the removed benzene volume is supplemented to 15 ml. The benzene is decanted to be free of water into a small separatory funnel and extraction is performed for 5 min with 2 ml of cobalt rhodanide reagent. After phase separation 10 ml of the benzene solution containing the surfactant-cobalt rhodanide complex is decanted into a centrifuge tube, an aliquot (7-8 ml) of the anhydorus benzene solution is pipetted into another separatory funnel and the same volume of water which has been previously acidified with 1 drop of concentrated HCl is added. The material is shaken for 5 min, the phases are allowed to separate and 5 ml of the aqueous phase, now containing the rhodanide, are pipetted into a 10 ml graduated flask and the rhodanide content is determined according to Bark and Higson. The calibration curve is constructed with 0-20 $\mu$g of pure substance of the nonionic surfactant investigated.

Cobalt rhodanide reagent: 20 g $NH_4CNS$, 3 g $Co(NO_3)_2 \cdot 6\ H_2O$, 25 g NaCl to 100 ml water. The solution is extracted with petroleum ether and the petroleum ether layer is discarded.

If the water sample also contains anionic surfactants, a suitable anion exchanger (for example, De-Acidite F.F.I.P. 100-200 mesh) is added to the foaming tube per 100 ml of water sample before the benzene addition, the tube is aerated for 5 min, during which the agitated exchanger binds the anionic surfactants. This is followed by alkalinization, addition of benzene and continuing the procedure as described above.

*Rhodanide determination according to Bark and Higson*[19]. The water sample of 5 ml or the hydrochloric acid solution after decomposition of the surfactant-cobalt rhodanide complex are treated with 1 drop of freshly prepared saturated bromine water and mixed in a 10 ml test tube. After standing for 1-2 min, the bromine excess is removed by shaking with 0.2 ml 2% $As_2O_3$ solution, 4 ml of freshly mixed pyridine-p-phenylenediamine hydro-chloride solution are added, the solution is brought to 10 ml with water, mixed, allowed to stand for 25 min and is read against water in the spectrophotometer at 515 nm in a 1 cm cuvette. The absorbance of a reagent blank value must be subtracted. The latter should not exceed $E = 0.05$.

Pyridine solution: 31 ml concentrated HCl are added to a mixture of 18 ml pyridine and 12 ml water.

p-Phenylenediamine dihydrochloride solution: 0.36 g of solid substance is diluted with water to 100 ml.

Reagent mixture: 3 ml pyridine solution + 1 ml phenylenedia-mine solution.

*Wickbold method*[335a]. While an older method (Wickbold[334]) involved the extraction of nonionic surfactants from an alkaline bicarbonate solution with butanol, dissolving the evaporation residue in methanol and water, followed by centrifuging, Wick-bold[335a] has recently also been using a foaming method. The evaporation residue of the organic phase is dissolved, is precipi-tated with barium-tungsten iodide (Dragendorff reagent), and the washed precipitate is dissolved in tartrate solution. Its bismuth content is determined with dithiocarbamate solution; in the older procedure, this is done by spectrophotometry at 405 nm, and re-cently by potentiometric titration.

For foaming off the surfactant in ethylacetate, an apparatus with a glass frit and drain cock is used as shown in Fig. 10.4. It serves for 1 l of water sample, scaled up also for 4-5 l with a surfactant content of up to 15 ppm. A moderate nitrogen stream is passed through for 2 x 5 min, each time with the addition of 200 ml ethylacetate. The filtered organic phase is evaporated, the residue is taken up in 5 ml methanol and 40-50 ml water, adjusted to pH 4-6, precipitated with 30 ml of precipitant and stirred for about 10 min.

For the photometric determination, the precipitate is centri-fuged, and after decanting of the clear supernatant, the residue is washed with 5 ml glacial acetic acid and centrifuged once more. The glacial acetic acid is then decanted, the precipitate is dis-solved in 3 ml tartrate solution and treated with 12 ml of dithio-carbamate solution in acetone. The absorbance is read at 405 nm. For a blank value 3 ml tartrate solution and 12 ml dithiocar-bamate solution are used. A suitable calibration line is con-structed with 30-300 $\mu$g Marlophen 810.

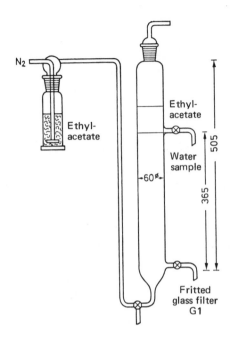

Fig. 10.4—Apparatus for de-foaming surfactants.

For a potentiometric titration the precipitate obtained with the Dargendorff reagent is filtered through a small fine-pored porcelain filter (A 2), washed with glacial acetic acid, dissolved in hot ammonium tartrate solution and diluted to 150-200 ml with water. The pH is adjusted to 4-6 and standard acetate buffer is added. A platinum-calomel couple is introduced and potentiometric titration is carried out with $N/2000$ or $N/4000$ pyrrolidinedithiocarbamate solution with stirring (magnetic stirrer) up to the potential jump. A blank value must be subtracted.

The calibration factor for Marlophen 810 amounts to 27 $\mu$g surfactant per ml $N/4000$ standard solution. The titer of the dithiocarbamate solution can be verified with a titer-stable $N/2000$ $CuSO_4$ solution.

Reagents: Precipitation reagent, solution A: 1.7 g basic bismuth subnitrate ($BiO \cdot NO_3 \cdot H_2O$) are dissolved in 20 ml glacial acetic acid and brought to 100 ml with water.

Solution B: 65 g potassium iodide are dissolved in 200 ml water. Solutions A and B are combined and after addition of 200 ml glacial acetic acid are diluted to one l with water. Two volumes of this mixture are mixed with 1 volume of a solution of 290 g $BaCl_2 \cdot 2 H_2O$ in 1 l of water. This percipitant is ready for use and is stable for 1 week.

Ammonium tartrate solution: 12.4 g tartaric acid and 17.6 ml 25% aqueous $NH_3$ solution are diluted with water to 1 l.

Pyrrolidinedithiocarbamate solution for photometry: 850 mg sodium pyrrolidinedithiocarbamate (Merck) are dissolved in 100 ml water, treated with 650 ml acetate and diluted to 1 l with water.

Photometer with Hg vapor lamp and Hg filter, 405 nm.

*Other methods for nonionic surfactant determination.* Wickbold[335] described the separation of a surfactant mixture by column chromatography on silica gel with butanone, silylation and gas-chromatographic separation; see Heinert[129] on thin-layer chromatography.

Williams and Graham[336] subject a residue containing nonionic surfactants to pyrolysis and use sodium nitroferricyanide to determine the forming aldehydes as a measure of the content of ethylene oxide derivatives.

See Lovell and Sebba[205] concerning a flotation method for surfactants.

The separation of ethylene oxide adducts after their acetylation by gas chromatography on columns with 4% Apiezon M on Chromosorb G, AW, silanized, at 380° has been described by Pollerberg[251].

Courtot-Coupex and Le Bihan[56] suggested extraction of the $Co(CNS)_2$ adducts with benzene and then to determine the cobalt by atomic absorption (see also Greff[116]).

Reed[256] described a thin-layer chromatographic separation of nonionics (silica gel layer, ethylacetate-glacial acetic acid solvent system, detection with a solution of bismuth oxynitrate and KJ in a mixture of $H_3PO_4$, dilute glacial acetic acid and $BaCl_2$).

The determination with a ring tensiometer after separation of the ionic surfactants with ion exchangers has been described by Koppe and Pittag[174].

## 10.5    Determination of Detergent Biodegradability

In order to relieve rivers from undegradable detergent residues from industry and homes which at one time had caused foam buildup often with a height of several meters and inhibited self-cleaning processes in water, West Germany enacted a regulation on 1 December 1962 (Bundesgesetzblatt 1962, No. 49, pp. 699-706) under which only those anionic detergents having a biodegradability of at least 80% are permitted. A test method and ap-

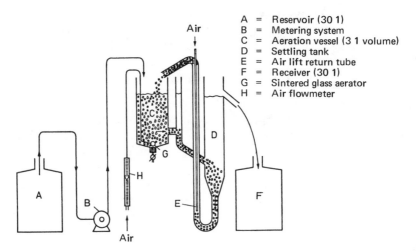

A = Reservoir (30 1)
B = Metering system
C = Aeration vessel (3 1 volume)
D = Settling tank
E = Air lift return tube
F = Receiver (30 1)
G = Sintered glass aerator
H = Air flowmeter

Fig. 10.5a.—Apparatus to test the biodegradability of detergents.

paratus, based essentially on the publications of Husmann, Malz and Jendreyko[143], were specified to determine this degradability.

Laboratory apparatus simulating the activated sludge process serves to circulate a synthetic waste water containing nutrient solution and 20 mg/1 surfactant (calculated as methylene blue-active substance [MBAS] according to the method in 10.2.1) with continuous aeration and recycling of the formed sludge, as well as continuous addition of fresh solution; the degradation obtained is determined by daily MBAS analyses.

See Fig. 10.5a for the experimental equipment. The reservoir **A** has a capacity for a least 24 1 (one day's ration for the prescribed feed of 1 1 synthetic waste water per hour); the receiver for treated solution has the same capacity.

Nutrient solution: 3.75 g peptone from caseine, 2.50 g meat extract, 0.650 g urea, 150 mg NaCl, 100 mg $CaCl_2 \cdot 2\ H_2O$ and 50 mg $MgSO_4 \cdot 7\ H_2O$ are dissolved in 1 1 of drinking water (EDITOR'S NOTE: Many equally good synthetic substrates are described in the literature).

Detergent solution: Sufficient surfactant or detergent is dissolved in 1 1 of distilled water to correspond to 9.60 g MBAS.

Synthetic waste water: 1 1 nutrient solution and 50 ml detergent solution are diluted to 24 1 with tap water (*i.e.*, a one-day ration). Thus, it contains 20 mg MBAS/1.

No bacterial inoculations are made for enrichment; the natural infection from the environment is considered sufficient.

Operation of equipment: The air feed is adjusted so that thorough mixing but no spattering or foam overflow occurs and a minimum dissolved oxygen content of 2 mg is maintained in the waste water. The feed of fresh and discharge of treated solution is adjusted to 1 l/hr. The Mammut air lift return sludge pump is to provide a uniform recycling of activated sludge from the settling vessel into the aerating vessel C. After 24 hr, the receiver contents are well mixed and a sample is taken for the MBAS determination according to 10.2.1. The degradation in per cent is calculated from the found MBAS content referred to the content in the fresh waste water sample.

The degradability is obtained as the arithmetic mean from the degradation values resulting after the end of the stabilization period on 21 consecutive days with approximately equal degradation values and defect-free operation of the equipment (see Fig. 10.5b). For a control of the nutrient degradation the $KMnO_4$ consumption* is determined every second day according to 5.4.1.1; its decrease as a criterion of nutrient degradation should show a continuous course. The organic fraction in the activated sludge solids should not exceed 3 g/l of waste water; otherwise the excess is to be removed.

Further details can be learned from the cited legislation and the publication of Husmann, Malz and Jendreyko.

Fischer[86] as well as Heinz and Fischer[130] and Janicke[147] report on practical experiences with the detergent test.

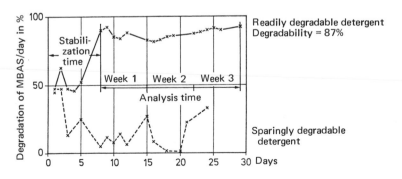

Fig. 10.5b.—Degradation curves of detergents.

---

*For U.S. equivalent see also dichromate consumption, Section 5.5.

# 11. NITRILOTRIACETIC ACID

Nitrilotriacetic acid salts have recently been proposed in place of polyphosphates as detergent additives in order to prevent the eutrophication phenomena caused by phosphates (see 1.3.3). Colorimetric and polarographic methods are available for their determination in water samples.

*Colorimetric determination according to Thompson and Duthrie*[311]. A water sample of 25 ml, together with a blank determination and a calibration determination—the latter with 5 mg NTA/l—is stirred with 2.5 g cation exchanger (for example, Dowex 50 W-X 8, 100-200 mesh) for 15 min. The sample is filtered without wetting or rinsing the filter. Of the filtrate 15 ml are mixed with 35 ml zinc-Zincon reagent. The absorbance difference is read at 620 nm against the blank solution in a 2 cm cuvette.

Calculation:

$$\text{mg NTA/l} = \frac{\text{blank absorbance} - \text{sample absorbance}}{\text{blank absorbance} - 5 \text{ mg standard absorbance}} \cdot 5$$

Preparation of ion exchanger: 100 g exchanger are stirred with 40 ml 6 N NaOH and 50 ml water for 10 min, decanted, stirred once more with 25 ml NaOH and 50 ml water in the same manner, decanted, washed 5-10 times with water and decanted until the wash water has pH values of 7.0-7.5. The exchanger is aspirated and aerated for 20 min.

Zn-Zincon reagents: Buffer solution: 31 g boric acid with 37 g KCl are dissolved in 800 ml water, adjusted to pH 9.2 with 6 N NaOH and diluted to 1 l.

Zinc solution: 0.440 g $ZnSO_4 \cdot 7 H_2O$ are dissolved in 100 ml 2 N HCl and diluted to 1 l with water.

Zinc-Zincon reagents: 0.130 g Zincon powder (2-carboxy-2′-hydroxy-5′-sulfoformazylbenzene Na-salt, Merck as well as LaMotte Chemicals, Chestertown Md.) is dissolved in 2 ml N NaOH, mixed with 300 ml buffer solution and 15 ml zinc solution in a 1 l graduated flask and diluted to 1 l. It is possible to analyze 0.2-10 mg $Na_3$ NTA/l; the standard deviation amounts to $\pm$ 0.2 mg/l.

For titration with Pb solution at pH = 4.4 against dithizon or with Fe(III) solution at pH = 3.5 against Tiron, see Fuhrmann, Latimer and Bishop[98].

Another colorimetric method for NTA determination has been described by Swisher, Crutchfield and Caldwell[303].

In a polarographic method for NTA determination which can be applied even in the presence of large amounts of ethylenediaminetetraacetate according to Daniel and Le Blanc[61a] (see also Farrow and Hill[80]), a small excess of cadium salt is added to the sample. During polarography at pH = 8 the wave corresponding to the Cd-NTA complex appears before that of cadmium. A procedure based on the enrichment method with the addition of graduated known NTA quantities has been described by Wernet and Wahl[331] (see also Haberman[120]).

# 12. HYDROCARBON DETERMINATION

## 12.1 Introduction

*Harmful and toxic effects; summary of aanlytical methods.* One of the most threatening and widespread hazards for our water supplies is caused by crude and heating oils as well as fuels and lubricants which introduce pollutants into surface and even ground water by accident or illegal storage and handling in their transport by motor vehicles and pipelines as well as other manipulations, for example, their use in gas stations, automotive repair shops as well as quite generally in workshops and production plants and residential oil heating systems.

Such pollutants can usually already be detected externally by iridescent oil films, tubidity or by their odor and taste.

The isolation of such hydrocarbon mixtures from water samples as a rule takes place by physical enrichment methods, such as extraction with solvents for fats, activated carbon, adsorptive precipitation or head-space distillation. However, these procedures isolate not only the hydrocarbons but also other materials with similar physical properties, for example, natural fats and waxes, plasticizers and chlorinated hydrocarbons, the presence of which must be subjected to a different evaluation and which must therefore be separated, usually by chromatography.

In contrast, in an analytical hydrocarbon determination it is rarely necessary—and often impossible—to separate the components down to the individual substances. It is usually sufficient to distinguish between light naphthas, oils and high-boiling residues on the basis of the evaporation properties. Occasionally the aromatics content (benzene, toulene, xylene) may be of interest because of the increased toxicity and taste effect of these compounds.

First it is necessary to distinguish between light naphthas, oils and high-boiling residues on the basis of their volatility and vaporizability. This finding furnishes information of origin as well as possibilities for eliminating the pollutants.

A quality reduction of water by mineral oil products primarily occurs as a result of an unpleasant, odor and taste. Two factors are important in this regard: on one hand, the water solubility of the hydrocarbons involved, and on the other, their specific odor and taste intensity, as a rule characterized by the threshold odor number (see 4.1), *i.e.,* the dilution at which an odor perception begins.

Table 12.1a shows the water solubility of some hydrocarbons. It decreases in the sequence of aromatics-naphthenes-paraffins and with increasing molecular weight. In the case of mixtures which are usually involved, for example, motor vehicle gasoline, the content of aromatics as well as their boiling points will be of decisive importance. The solubility of hydrocarbons is significantly increased by the presence of solubilizers, for example, detergents.

Table 12.1a.—Water solubility of hydrocarbons in mg/l
(according to Kempf[163] and Fastabend[81])

|  |  |  |
|---|---|---|
| n-Hexane |  |  |
| Cyclohexane |  | 60 |
| Benzene |  | 1680 |
| Toluene |  | 511 |
| Petroleum spirits |  | 31 |
| Automotive gasoline | I | 361 |
|  | II | 146 |
|  | III | 93 |
|  | IV | 505 |
| Diesel fuel |  | 17.22 |

The individual hydrocarbon groups differ particularly in their odor intensity; in this regard pure n-paraffins are least perceptible. The more volatile homologs, such as n-heptane to n-nonane have a rather pleasant fruity odor, while the higher-boiling compounds, for example, high-melting wax or pharmaceutical paraffin oil, are odorless. The typical gasoline odor is produced primarily by isoparaffins and naphthenes, but particularly by aromatics; thus, coal tar products with a high aromatics content, for example, naphthalene, are of outstanding importance with regard to their influence on odor and taste.

Because the compositions are not exactly specified, literature data on the threshold odor numbers of industrial-grade mixtures, such as "automotive gasoline" or "mineral oil" in water fluctuate by several orders of magnitude; moreover, consideration must be given to the varying individual sensitivity of subjects evaluating the odor of a water sample (see 4.1).

A toxicological evaluation of water pollutants by "hydrocarbons" also cannot be made on the basis of overall data; rather, at least the quantitative ratio of the above-cited groups of compounds must be taken into account. Little is known on the toxic effects of hydrocarbons in concentrations that may occur in drinking water after accidents, for example. If we compare the maximum permissible concentrations (MPC) of some typical individual hydrocarbons in the atmosphere of a working site as specified by law (see, for example, Leithe[187]), we recognize the higher toxicity of aromatics compared to the paraffins (see Table 12.1b).

Table 12.1b.—MPC values in mg/m³ of air for 1966

| n-Hexane | 1800 | Gasoline | 2000 |
|---|---|---|---|
| n-Octane | 2350 | Benzene | 80 |
| Cyclohexane | 1050 | Xylene | 870 |
| | | Naphthalene | 50 |

Since a working man inhales at least 2-3 m³ of air during an 8-hour day, he absorbs a total quantity of hydrocarbons which is hardly ever reached in dissolved form in polluted drinking water. Nevertheless, the public health expert will justifiably designate a water as unfit when it shows a marked odor of gasoline, petroleum or tar oils. According to the Food Act such water is impotable.

The toxic effect of dissolved hydrocarbons on water life, for example, fish, is also highest in the case of aromatics (critical concentration: 5 mg/l for benzene, 50-200 mg/l for gasoline depending on grade, 2.5-5 mg/l for naphthalene; see Zahner[344] as well as Meinck, Stoof and Kohlschütter[218]). In the case of diesel, automotive and fuel oils, toxicity limits of 50-100 mg/l have been reported; however, with such concentrations these are no longer present in dissolved form. Paraffin hydrocarbons are relatively easily digested by bacteria and converted into biomass, a process which is known to be under consideration for the recovery of nutrients and feeds from petroleum.

The hydrocarbon concentration of extracts obtained with organic solvents can be determined by various methods. If a volatile hydrocarbon determination is not required, the low-boiling solvents (petroleum ether, chloroform) can be evaporated and the residue weighed; in a subsequent procedure, for example, by adsorption on clay or Florisil, the "polar" fractions, which do not consist of inert hydrocarbons, for example, natural fats, can be separated from the residue.

With the use of high-purity solvents with well-defined properties differing sufficiently from those of hydrocarbons, binary mixtures are obtained and their composition can be determined by suitable physical tests. Thus, with the use of a solvent of high specific gravity a density determination in the Pyknometer can furnish the content of the hydrocarbons of lower specific gravity. Such analyses have been described, making use of carbon tetrachloride and, more advantageously, with tetrabromoethane which has a particularly high density.

Hydrocarbon solutions in carbon tetrachloride can be analyzed by IR-spectrophotometry at a wavelength specific for CH-bonds where the solvent is largely transparent.

For a semi-quantitative determination of higher hydrocarbons their fluorescence can also be utilized.

If volatile hydrocarbons are to be determined exclusively, for example, gasoline or petroleum, they can be evaporated from the water sample. The hydrocarbon content of the vapors can be determined by combustion into $CO_2$ in the course of elemental analysis, by introducing a gas test tube specifically for hydrocarbons or by adsorbing the hydrocarbons on activated carbon with a determination of its weight increase, or by turbidity reactions.

Hydrocarbon extracts can be separated further by TLC and detected with suitable color reactions.

Numerous procedures have also been published on the separation and determination of hydrocarbons by GC in which the nonvolatile components are subjected to prepyrolysis.

## 12.2 DEV Gravimetric Method, H 17/18 (Old Edition)

This method, which has been replaced by more recent data (see below), can still find occasional use, for example, as a part of precipitation experiments.

Principle: the hydrocarbons are adsorbed from the water sample by entrainment-precipitation on aluminum hydroxide, are extracted from the latter with petroleum ether and after evaporation of the latter are weighed. In addition to hydrocarbons the extract also contains petroleum ether-soluble materials, for example, natural fats and oils, plasticizers, chlorinated insecticides, etc. Depending on the procedure used for drying the residue, the volatile fractions are lost.

Procedure: One ml aluminum sulfate solution is added to each liter of water sample and after shaking, 1 ml $Na_2CO_3$ solution is added. The precipitate, which contains the oily water components in adsorbed form, is allowed to settle, the supernatant water is decanted, the precipitate is dissolved with a small amount of hydrochloric acid, followed by repeated extraction with petroleum ether. The petroleum ether extracts are washed repeatedly with water and dried for 12 h on anhydrous sodium sulfate. The petroleum ether is evaporated on a water bath and the residue is weighed.

Aluminum sulfate solution: 30 g $Al_2(SO_4)_3 \cdot 18 \ H_2O$ and 70 ml water.

$Na_2CO_3$ solution: 20 g $Na_2CO_3 \cdot 10 \ H_2O$ and 80 ml water.

Burmeister[45a] has proposed the following more exact procedure:

One l of water sample is adjusted to pH = 5 with hydrochloric acid. Two ml of the above 30% aluminum sulfate solution are added and the pH is adjusted to 7 with 2 ml of 20% $Na_2CO_3$ solution. The precipitate formed is allowed to stand for 12 h, is filtered through black-ribbon filter paper, rinsed into a separatory funnel with 200 ml water, followed by an addition of 20 ml 10% hydrochloric acid and 3 extractions with 50 ml portions of petroleum ether. The combined petroleum ether extracts are washed 3 times with 50 ml portions of water until neutral, are dried on anhydrous sodium sulfate for 12 h and filtered on blue-ribbon filter paper. Sodium sulfate and the filters are reextracted 3 times with 5 ml portions of petroleum ether and the petroleum ether extracts are concentrated at 80° in a weighed 50 ml round-bottom flask. The residue is dried for 1 h at 80° and weighed.

A separation into "non-polar" hydrocarbons and "polar" natural fats and the like can be realized by passing the petroleum ether solution through a 12 cm column of $Al_2O_3$ (activation stage 1, Merck 1077) and elution of the hydrocarbons with 50 ml petroleum ether. Concentration is performed as above at 80°, followed by drying and weighing.

*Extraction with carbon tetrachloride and weighing of the evaporation residue.* Giebler, Koppe and Kempf[105] charge a 1 l separatory funnel with 10 ml NaOH solution (d = 1.11), 10 ml 10% Complexon-III solution (to prevent the formation of lime soap emulsions), 500 ml water sample, 50 g NaCl and 50 ml $CCl_4$. Shaking is allowed to take place for 1 h (shorter shaking periods have proved to be insufficient), the $CCl_4$ layer is largely eliminated by centrifuging, an aliquot is filtered and evaporated in vacuum (2 torr), followed by drying for 10 min and weighing.

*Gravimetric analysis of hydrocarbons boiling above 200° (760 torr) according to the method of the "Water and Mineral Oil" Committee as well as DEV H 17/18, 1971 Edition.* A water sample of 500 ml with 10 ml 10% NaOH solution, 10 ml 10% Complexon-III solution, 50 g NaCl and 5 ml $CCl_4$ is vigorously shaken for 1 h in a separatory funnel. After addition of 1 ml 25% sulfuric acid, the $CCl_4$ layer is centrifuged, filtered into a small graduated cylinder on a small cotton ball, and an aliquot of the filtrate (3.8-4.0 ml) in a weighed flask is concentrated in two steps in vacuum evaporation apparatus at a water bath temperature of 20° as described in the original publication. The first stage at 50 torr (water-jet vacuum) up to a volume of 0.5 ml requires 3-4 min, while the second stage at 2 torr (oil vacuum) also requires 3-4 min. Model evaporation experiments are carried out with 100 mg light fuel oil and 5 ml $CCl_4$. Under the cited conditions at least 95% of the original weighed sample should remain as a residue.

If only hydrocarbons without oxygen-containing constituents are to be determined, the centrifuged $CCl_4$ solution is shaken for 3 min with 0.5 g Florisil, filtered, and an aliquot of the filtrate is processed further.

Extraction can also be carried out by shaking for 2 min with 25 ml petroleum ether (b.p. 40-60°). The settled aqueous phase is extracted two more times with 15 ml portions of petroleum ether. The petroleum ether extracts are filtered, evaporated in the above-described vacuum apparatus and weighed.

*Activated carbon-chloroform extract (Carbon-chloroform extract, APHA CCE method)* Principle and range of validity: A large water sample (20 m³) is conducted over activated carbon. The carbon is dried and extracted with chloroform. The chloroform-free extract is weighed and can serve for various further identification procedures (see 3.2). The procedure does not allow a determination of all organic pollutants, for example, synthetic detergents, but the quantity of extract characterizes the quality of a given water to some extent. In the U.S. extract volumes of more than 200 µg/l are considered impermissible for the production of drinking water; clean surface and ground water contains only 25-50 µg/l extract.

The adsorption tube has a length of 45 cm and a diameter of 7.5 cm. An 11 cm section at the top and bottom is packed with 4.10 mesh activated carbon (Cliffs Co., Dow Chemical Co., Marquette, Mich.), while the center section is packed with 30 mesh

Nuchar c-190 of West Virginia Pulp and Paper Co., N.Y. Samples of these carbons in the packing as described should give a blank value of not more than 40 mg (4.10 mesh) or 20 mg (30 mesh) chloroform extract.

The water sample flow rate is adjusted to 1 l/min; 20 m³ of water sample are treated. Subsequently, the activated carbon is dried in clean air in an exposed layer at not more than 40° and is extracted for 35 h with chloroform in a Soxhlet extractor. The chloroform is evaporated completely and the residue is weighed.

*CAE method (carbon alcohol extract).* In the 13th edition of the APHA Standard Methods the CCE method was followed by an alcohol extraction. The above adsorption tube is packed only with 30-mesh activated carbon. The water sample is allowed to flow through slowly (120 ml/min, 1200 l per week). After the above chloroform extraction, the chloroform remaining in the adsorption tube is blown out with warm air and the carbon is extracted with 95% ethyl alcohol for 24 h. The alcohol is evaporated and the residue is dried at 75° until the weight decrease is less than 1% after a period of 72 h.

*Determination of volatile hydrocarbons by weighing on activated carbon.* According to Kempf[163] the water sample, charged into a suitably large Drechsel wash bottle with glass frit, is blown out at room temperature with a moderate bubble count of air which had been dried on soda lime the gas phase obtained is dried with $CaCl_2$ and magnesium perchlorate in a drying tower and conducted into a U-tube packed with activated carbon of 2-3 mm particle size (predried at 110° in an air stream); the weight increase of the tube is determined by weighing.

Goebgen and Brockmann[109] combust the blow-out hydrocarbons and determine the $CO_2$ formed.

## 12.3    Extraction Methods and Density Determination of the Solution

Similar to older pyknometric fat analysis in milk (Leithe[185]) in which the fat was taken up in carbon tetrachloride and the resulting density decrease served as a criterion of the fat content, a method has been described by Levine, Mapes and Roddy[202] in which the water sample is extracted with $CCl_4$ and the tetrachloride solution obtained is measured in a pyknometer. A limit of detection of about 10 mg/l has been reported (see also Rather et al.[255] as well as Headington et al.[126a]).

A significant increase in accuracy (limit of detection of 0.3-1 mg hydrocarbons/l) was realized by Ladenburg[180] by the use of 1,1,2,2-tetrabromoethane as the solvent which has a density of 2.956 (25°). This not only improves the sensitivity of the method but also reduces the errors resulting from density differences of the hydrocarbons present. A disadvantage of this solvent is its sensitivity to hydrolysis at a pH above 5.0.

Procedure: A 5 l water sample of pH 7 is treated with 50 ml buffer solution I and 10 ml buffer solution II to adjust a pH of 5.0-5.1. Exactly 20 ml of 1,1,2,2-tetrabromoethane are added and the mixture is stirred for 3 min in a mixer to form a stable emulsion. The mixture is treated with 5 ml aluminum sulfate solution in a separatory funnel. After standing for about 1 h, the supernatant turbid liquid is siphoned off and the sludge is centrifuged at 5000 rpm. About 15 ml of the lower liquid layer are pipetted into a flask, 3 g anhydrous magnesium sulfate are added, the mixture is shaken for 2 h and is allowed to settle and dry overnight. It is then filtered into a 10 ml capillary pyknometer, stabilized for 20 min to 25 ± 0.01° in a thermostat and weighed. A blank value is obtained in the same manner by shaking tetrabromoethane with water and then following the same treatment. It is advisable to prepare a tared flask with the blank weight of the filled pyknometer, so that only the weight difference of solvent and extraction solution needs to be determined in the analyses.

The hydrocarbon content is calculated on the basis of the mixing rule:

$$(V_a + V_b) \cdot D_{(a+b)} = V_a \cdot D_a + V_b \cdot D_b$$
$$(V = \text{volumes} \quad D = \text{densities})$$

In the present case, 18.9 ml of tetrabromoethane is use instead of a volume of 20.0 ml with consideration of the water-soluble amount. Since practically linear relationships exist in the expected low concentration range under simplified assumptions, the following formula results:

$$x = \frac{(P_1 - P_2) \cdot 18.9}{\left(\dfrac{d_{solv.}}{d_{oil}}\right) - 1} \cdot 10$$

x = mg hydrocarbons per 5 l of water sample

($P_1 - P_2$ = weight difference of pyknometer in mg.

A mean value of 0.88 was used for the hydrocarbon density ($d_{oil}$).

Buffer solution I : 822 g $NaH_2PO_4$ • $H_2O$, diluted to 1 l with water.

Buffer solution II: 83 g $Na_2HPO_4$ • 12 $H_2O$, diluted to 1 l with water.

Aluminum sulfate solution: 100 g $Al_2(SO_4)_3$, analytical grade, diluted to 1 l.

Anhydrous $MgSO_4$ dried at 400°, finely powdered and dried once more at 400°.

## 12.4    Determination According to Lawerenz[181] with Gas Test Tubes

Apparatus: A 1½-2 l three-neck flask with gas inlet tube, thermometer and attached reflux column with Raschig rings, followed by a $CaCl_2$ tube and a gas test tube KW 1 for gasoline vapors, and finally connected to a water-jet pump.

The water sample is heated to 70-90°; 2 l of air are passed through in 10 min with a moderate flow rate. The hydrocarbon content results from the evaluation of the colored zone corresponding to the manufacturer's instructions.

## 12.5    Determination of Volatile Hydrocarbons by Displacement and Turbidity Measurement (According to Sherrat[288])

Principle: Volatile hydrocarbons are displaced from the water sample by heating in an air stream, are adsorbed on activated carbon in a tube, extracted from the latter with acetone, precipitated with water and measured in the form of turbidity.

Apparatus: A round-bottom flask of 1.5 l capacity with gas inlet tube and reflux condenser. A glass tube of about 20 cm length and 5 mm inside diameter, packed with activated carbon of 0.35-0.75 mm particle size is attached on the condenser.

Procedure: The water sample is adjusted to pH 10 with sodium hydroxide and is heated to boiling in an air stream (the bubbles being just countable), while a condensate of 8-12 drops/min is maintained. Easily volatile hydrocarbons are displaced in 15-20 min and more sparingly volatile ones in 30-45 min.

The carbon filter is then eluted with acetone added in drops. When 2 ml acetone have flowed into a 10 ml graduated flask, they are treated with 8 ml Teepol solution. The turbidity formed is read in a suitable colorimeter at 490-550 nm or in a turbidimeter. The analysis is made on the basis of a calibration series of 0.2-1 ml of a 1% solution of the respective hydrocarbons (gasoline, kerosene) in acetone, diluted to 2 ml with acetone and treated with 8 ml Teepol solution.

Teepol solution: 1 ml Teepol (Shell) and 1 ml concentrated $H_2SO_4$ per liter of water.

For an analogous determination of kerosene (diesel oil) by turbidity measurements, Lee and Walden[183] extract 1 l of water sample for 5 min with 0.2 g activated carbon, filter on glass wool, dry briefly with a water-jet pump, wash the hydrocarbons 5 times with 3 ml portions of acetone, evaporate the acetone extracts to 2 ml, precipitate in a 10 ml graduated flask with a solution of 1 g Na laurylsulfate + 1 ml concentrated $H_2SO_4$ in 1 l of water, and read the turbidity as described above.

## 12.6 Detection of Mineral Oil Traces by Fluorescence

Data for the analytical determination of mineral oil traces in water samples on the basis of their fluorescence have been described by Nietsch[239a]. Concentrations of 0.1-1 mg/l can be detected directly under an analytical quartz lamp (UV—excitation at 366 nm) by a yellowish-white fluorescence. Even lower concentrations (starting at 1 $\mu$g/l) can be detected by the following method:

The water sample is shaken with very finely ground magnesium oxide (0.01 g MgO/100 ml water sample) and is filtered through a small paper filter. The filter is spread on a watch glass and examined in UV-light. It is also possible to extract 100 ml of water sample with 10 ml petroleum ether which is then evaporated.

The fluorescence color and intensity naturally are highly dependent on the composition of the mineral oil in question. Consequently, quantitative data can be obtained only by comparison measurements with a test mineral oil which is as similar as possible to that present in the sample.

Bauer and Driescher[19a] investigated the dependence of the fluorescence intensity as well as the energy distribution of the fluorescence spectra on the composition in different mineral oil products.

## 12.7    Thin-Layer Chromatography Determination by "Channel TLC"

(Method of the Water and Mineral Oil Committee as well as DEV H 17/18, 1971). Principle: The water sample is extracted with $CCl_4$, Aliquots of the $CCl_4$ solution are applied on a thin-layer plate divided into narrow development channels and subjected to ascending chromatography with $CHCl_3$. This results in a separation of the more strongly adsorbed oxygen-containing impurities. The hydrocarbons are determined in a common zone and analyzed.

Five tapered channels of 2 mm width and 50 mm length are formed by scoring a ready-to-use silica gel TLC plate (Merck $F_{254}$) with a metal stylus and a template (Fig. 12.7).

One l of the water sample is treated with 10 g NaCl and is extracted with 2 ml $CCl_4$ for 10 min to 1 h by vigorous shaking. The $CCl_4$ phase is discharged into a 10 ml upright cylinder with a small amount of water. With the use of a 100 μl Hamilton syringe, 20 μl of tetrachloride solution are applied into the funnel-shaped flares of channels 1 and 2, while 100 μl of solution are introduced into channels 3-5. After evaporation of the solvent, the plates are developed in a conventional thin-layer apparatus with chloro-

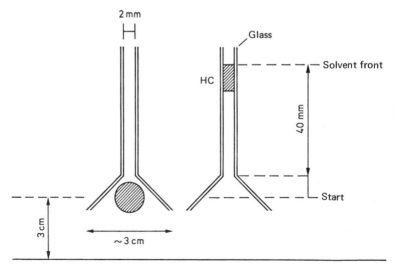

Fig. 12.7.—Channel thin-layer chromatography; thin-layer plate.

form as the solvent until the solvent front has moved through 40 mm of the channels. This is followed by air drying and the plates are then placed into a glass tank saturated with iodine vapors. The hydrocarbons appear in the form of rectangular brown spots; their area is carefully measured.

The procedure is calibrated with a standard solution of 250 mg fuel oil in 50 ml $CCl_4$ under the same operating conditions and the area constant in $mm^2$ per mg of hydrocarbons is calculated. The lower limit of detection is reported to be 0.5-1 ppm hydrocarbons. The sensitivity can be increased by concentrating the $CCl_4$ solution or with the use of more narrow channels.

A maximum error of $\pm$ 0.5 ppm was found in the 2-15 ppm range.

*Other TLC methods for hydrocarbons.* Giebler, Koppe and Kempf[105] conduct the TLC separation of the $CCl_4$ extract, concentrated if necessary (see p. 149) by applying it 1 cm from the lower edge of 20 x 20 cm plates (0.25 mm layer of a mixture of equal parts of aluminum oxide G and kieselguhr 9), drying in air and chromatographing in the ascending direction in hexane to a development distance of about 10 cm in 10 min. The plate is air-dried for 5 min and is sprayed with 7 ml of a 1% solution of phosphomolybdic acid in tert.-butylalcohol. It is allowed to dry in air for 15 minutes and treated in a drying oven at 120° for 30 min. The hydrocarbons appear with a blue-green color on the solvent front, while "polar" components are brown-gray at the start.

Rübelt[263] allows 100 l of water sample to flow at a rate of 5 l/h to a tube of 2 cm diameter packed with activated carbon (0.3-0.8 mm) with a filling height of 12 cm. The wet carbon is extracted with carbon tetrachloride in a flask with a water separator and reflux condenser for 1 h. For a direct extraction of the water sample it can be stirred with 25 ml $CCl_4$ for about 1 min with the use of a high-speed blade agitator (3000-4000 rpm).

Separation of the "polar" fractions can be carried out with 2 g $Al_2O_3$ (activity stage 1) on a column of 7 mm diameter or by shaking for 2 min with 1 g Florisil (60-80 mesh).

Thin-layer chromatography is carried out on silica gel H, activated for 1 h at 110°, layer thickness 0.3 mm, with n-hexane as solvent. The components are detected by spraying with 0.03% uranine solution and observation in UV-light of 366 nm. The Rf-values range between 0.88 and 1.

Dietz and Koppe[63] detect the aromatics in UV-light without a spray. Subsequently, they spray with rhodanine B solution and examine the non-aromatic hydrocarbons in UV light.

Goebgen[108a] uses silica gel F-254 (Merck ready-to-use plates) and n-hexane as the solvent. A spray reagent of 50 mg bromthymol blue, 1.25 g boric acid, 8 ml N NaOH in 112 ml water serves for detection. Yellow spots form on a blue background (development time 15 min, Rf-values of the paraffinic hydrocarbons about 0.6; limit of detection 2 $\mu$g; see also Berthold[25a]).

*Determination by microcircular TLC (MCTLC) according to Koppe and Muhle*[173]. Koppe and Muhle[173] extract 3 l of water sample with 12 ml $CCl_4$ for 2 h and dry the $CCl_4$ extract on $Na_2SO_4$. After removal of the polar components on $Al_2O_3$ the $CCl_4$ solution is subjected to MCTLC (see 14.8). Detection is obtained by spraying with an 0.5% phosphomolybdic acid solution in isobutanol and heating to 80°.

Another fraction of the $CCl_4$ solution can be brominated by boiling with a bromine solution in $CCl_4$ with irradiation and can also be detected by MCTLC.

## 12.8    GC with Direct Injection

The gas-chromatographic determination of hydrocarbons by direct injection of the water sample is rapid and convenient because the isolation procedures are eliminated; however, it has the disadvantage that the FID detection sensitivity is greatly reduced by large quantities of water. For the gas-chroatographic analysis of industrial effluents containing a mixture of numerous organic pollutants, see Sugar and Conway[302]. These authors use a column (5 m x 3 mm) packed with 5% Carbowax on Chromosorb W with temperature programming of 4°/min between 50 and 250°. IR-analysis and mass spectrometry serves for peak identification.

Jeltes and Veldink[150] determine small gasoline quantities in water after extraction with nitrobenzene. The 3 m column is packed with 10% polyethyleneglycol 1500 on silanized Chromosorb W (60-80 mesh). The temperatures are 68° in the column, 217° on the injector and 220° in the FID. Carrier gas: $N_2$, 25 ml/min. The alkanes emerge as a common peak separate from the aromatics. Nitrobenzene is eluted considerably later.

According to Jeltes[149] higher-boiling hydrocarbons are extracted with $CCl_4$ and concentrated, followed by GC on a 1.8 m column (3 mm I.D.) packed with 5% SE 30 on Chromosorb W with temperature programming from 85 to 300° (temperature in the vaporizer block 350° and in the detector 370°).

See also Banyon, Kaschnitz and Rijnders[18a] on the GC determination of volatile and higher-boiling hydrocarbons.

*n-Paraffin isolation and determination.* n-Paraffins of natural origin also are present in small concentrations in surface water which is not polluted by waste effluents. Rübelt[263] makes use of the urea adducts on thin-layer plates for their isolation and determination.

For preparation of the plate material 50 g silica gel G are slurried in 90 ml of an aqueous solution containing 15 g urea, applied in a thickness of 0.5 mm on 5 plates (20 x 20 cm) and dried at 70° in an oven.

Above the application line the n-paraffins form inclusion compounds which are detected with uranine solution (see above), scraped off, extracted with $CS_2$ and separated into individual n-paraffins by gas chromatography.

This is performed in a 1.5 m column of 3 mm diameter packed with 1% SE 30 on Chromosorb W, 60-80 mesh. Temperatures: injection block 280°, FID furnace 350°; carrier gas 30 ml $N_2$/min; FID 20 ml $H_2$, 70 ml $O_2$/min.

## 12.9 Hydrocarbon Determination by IR-Analysis

Simard, Hasegawa, Bandamk and Headington[290] provide the following procedure: A water sample of 3 l is extracted with 20 ml carbon tetrachloride for 15 min in an efficient agitating machine, preferably in the glass vessel serving for sampling. Of the $CCl_4$ solution 15 ml are filtered into the cell of an IR-spectrophotometer with NaCl-prism and read in the region of 3.42-3.50 $\mu$m.

A mixture of 37.5% isooctane, 37.5% Cetan and 25% benzene serves for calibration. The sum of the 3.42 and 3.50 $\mu$m absorbances is plotted as a function of 5-100 ppm concentrations in $CCl_4$.

Measurements of the authors showed that waste oils from the most diverse production processes show between 74 and 126% of the theoretical values. A xylene mixture resulted in only 18% of theory (see also Fastabend[81]).

Ludzak and Whitfield[208a] adjust 1 l of water sample to a pH of less than 3.5 with HCl and extract it in a liquid-liquid extractor with $CCl_4$ for 4 h; the extract is concentrated to 10 ml, dried with anhydrous $Na_2SO_4$, adjusted to a suitable volume and the IR-absorption is read at 2925 cm⁻¹ ($= 3.42$ $\mu$m).

For the separation of components which are not only hydrocarbons (e.g., natural fats and oils), the $CCl_4$ solution is concentrated to 2-3 ml, loaded on a column with activated $Al_2O_3$ and eluted with $CCl_4$; 30 ml of eluate are concentrated and read in the IR instrument as above.

See also Unger[317], Hellmann[132] as well as Rübelt[262].

*IR-method according to the procedure of the "Water and Mineral Oil" Committee and DEV, H 17/18 (1971).* A water sample of 1 l in a square bottle of 1.5 l capacity is stirred with 25 ml $CCl_4$ for 30 sec at 3000-4000 rpm. The settled $CCl_4$-layer, centrifuged if necessary, is dried on anhydrous $Na_2SO_4$, filtered, extracted for 2 min with 1 g Florisil (60-100 mesh) to separate non-hydrocarbons, and filtered through quartz cuvettes of 5, 10 or 50 mm optical path. The reading is taken in a IR-spectrophotometer against pure $CCl_4$. An adjustment to 100% transmission is made at 3.2 $\mu$m and the spectrum is recorded at 3.2-3.6 $\mu$m. The absorbance at 3.30, 3.38 and 3.42 $\mu$m is determined from the spectra (see Fig. 12.9).

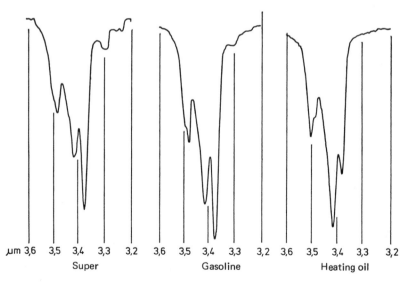

| μm 3,6 | 3,5 | 3,4 | 3,3 | 3,2 | 3,6 | 3,5 | 3,4 | 3,3 | 3,2 | 3,6 | 3,5 | 3,4 | 3,3 | 3,2 |

Super    Gasoline    Heating oil

Fig. 12.9.—IR-spectra of different mineral oil products in the range of the CH-valence vibrations.

The following calculating formula has been developed for carburetor fuels (CF):

$$c_{CF} = \frac{1.3 \cdot b}{a \cdot d} \; (1.1 \, E_{3.30} + 0.12 \, E_{3.38} + 0.19 \, E_{3.42})$$

while the formula below applies to all mineral oil products except tar oils:

$$c_{oil} = \frac{1.4 \cdot b}{a \cdot d} \; (0.12 \, E_{3.38} + 0.19 \, E_{3.42}).$$

$c$ = CF or oil concentration in mg/l,
$a$ = volume of water sample in l,
$b$ = volume of $CCl_4$ in ml,
$d$ = optical path in cm,
$E$ = absorbances.

In these formulas the relative frequency of the $CH_2$ and $CH_3$ groups responsible for the valence vibrations is taken into account.

The limit of detection is reported to be 0.05 mg/l for instruments without scale expansion. For oils with a high aromatics content the more sensitive UV-fluorescence method is recommended. Higher accuracies can be realized with calibration curves constructed with the use of suitable hydrocarbon mixtures.

# 13. DETERMINATION OF VOLATILE CHLORINATED HYDROCARBONS

Volatile chlorinated hydrocarbons, for example, trichloroethylene and carbon tetrachloride, are frequently used for dry cleaning of garments, degreasing of metal parts as well as linseed extraction, and can therefore reach waste water effluents. They are preferably determined by means of GC either by head space analysis (see 6.4.3) or in a hexane extract; such chlorinated hydrocarbons are indicated with particular sensitivity by an electron capture detector.

A simpler procedure is to determine the components by colorimetry with pyridine and sodium hydroxide solution on the basis of the Fujiwara reaction. Thielemann[310] analyzes waste water for trichloroethylene as follows:

A water sample of 5 ml with 9 ml pyridine and 5 ml 50% sodium hydroxide solution is immersed for 1 min into a water bath heated to 70-80°, diluted with 3 ml hot water, and the blue-red color of the pyridine layer is immediately read at 430 nm in the photometer. A calibration curve is constructed with 0.1-100 mg trichloroethylene/l.

*Carbon tetrachloride determination according to Stankovic[296a].* A water sample of 10 ml in a glass-stoppered test tube is shaken vigorously with 10 ml methylethylketone for 5 min. After layer separation 5 ml pyridine, 1 ml of the upper methylethylketone layer and 2 ml 20% sodium hydroxide solution are shaken in a second test tube, heated for 3 min in a boiling water bath and quenched in cold water. A sample of 5 ml is pipetted from the pink layer, treated with 3 ml ethanol and immediately read in the photometer against a blank experiment with distilled water in a 1 cm cuvette with a S 53 filter. A calibration curve is constructed with solutions of 0-16 mg $CCl_4$/l. Five mg $CCl_4$/l produced an absorbance of 0.34; even 0.09 mg $CCl_4$/l can be detected.

# 14. PESTICIDES

## 14.1 Introduction: Harmful Effect; Toxicity

Pesticides, particularly repellants of the most diverse type, are finding continuously more widespread use in agriculture and forestry, horticulture and stock-piling. This results in increased hazards of pollution of surface and ground water with such agents, some of which are extremely potent, particularly when instructions for use are not carefully observed. Although many pesticides, particularly the chlorinated hydrocarbons, are only very sparingly water-soluble, this solubility suffices to have an influence on the taste of ground water, for example, or to cause fish deaths in surface water. The solubility can be notably increased, for example, by the presence of surfactants (detergents). The water analyst will therefore be increasingly faced with the task of identifying and quantitatively determining such materials in water, often in extremely low concentrations.

Although it would be desirable to detect such toxins in water even before damage has been produced and to take prompt preventive measures, the analyst is usually called only after fish or domestic animals have been poisoned in order to discover the reason for such events.

The procedure in such problems will largely depend on existing observations. Frequently, a noticeable odor or pathological diagnosis in animals will furnish information on the presence of certain toxins. When facilities are available for biological experiments and breeding the necessary experimental plant or animal organisms, testing of the biological effect in many cases can not only confirm suspicions or findings concerning damage that has occurred but can also simplify the selection of a method of chemical analysis.

Chemical analysis methods for traces of repellants in water have been greatly enriched by new methods of residue analysis in commercial plant and animal products; this field is under intense analytical study because of the existence of intoxication hazards and frequent expressions of alarm, so that a very abundant literature is available on the subject to which the water analyst can refer. Apparatus for the purpose has become commercially available in wide selection and is equipped with auxiliary devices and instructions for pesticide analysis by the manufacturers.

The analysis of repellants is made more difficult by the fact that there is hardly another field of application for chemicals with such a diversity of active ingredients. On one hand, this results from the need to continually reduce the possible hazards for man and domestic animals and plants; on the other hand, the capability of many pests to develop resistance strains which no longer react to previously effective toxins requires the development and marketing of new agents, often belonging to very different chemical groups of compounds.

The following compound groups can be listed as the most important at the present time:

1. Chlorinated hydrocarbons as insecticides,

2. Halogenated phenoxy-acid derivatives as weed killers (selective weed control agents),

3. Organic phosphorus compounds, primarily in the form of quick-acting or systemic insecticides,

4. Carbamates as insecticides and fungicides.

*Pollution of water by pesticides.* With regard to water pollution by pesticides, hazards for drinking water must be considered in the first place. These increase with water solubility, toxicity and stability of the agents involved. Beran and Guth[24] have made reference to the high retention power of some soils (except for light sandy soils) for commercial pesticides, allowing them to enter spring and ground water, for example, as a result of their application over large areas. On the other hand, some chlorinated hydrocarbon insecticides have a life of several years in soil. In any case, the hazard of intoxication of man by pesticides through drinking water is considerably smaller than through residues on food treated with them. Moreover, the selection of applied pesticides is increasingly based on active ingredients with the minimum toxicity for man or with a lower stability.

The situation in *surface water* is considerably more critical. The possibility of pollution by pesticides always exists with their incorrect application whether they are used over large areas or whether spray and dusting agents are handled carelessly. Although there is hardly any hazard for human intoxication, for example, in swimming, fish and other water life serving as a basic nutrient for fish are particularly endangered because of the high toxicity of many of these agents for water fauna in particular.

Because of the high stability of some chlorinated hydrocarbons, these are enriched in some organisms or organs which then serve as food for other animals and finally for man, reaching concentrations which give rise to serious concern among public health specialists.

Table 14.1 cites some data which are of interest for water conservation with regard to pesticides and their unfavorable

Table 14.1.—Data on the toxicity of pesticides

| Active ingredient | Water solubility (mg/l) | MPC mg/m$^3$ air | LD$_{50}$-Rat mg/kg body weight | LD$_{50}$-fish mg/l water |
|---|---|---|---|---|
| Aldrin | 0.03 | 0.25 | 67 | 0.2 |
| Captan | 0.5 | | 9000-15000 | |
| Carbaryl | Less than 1000 | 5 | 400-850 | |
| Chlordane | | 0.5 | 250 | 0.05-1 |
| Dalapon | Readily soluble | | 6600-8100 | 340 |
| Dichlorvos (Vapona) | 10000 | 1 | 50-80 | 1000 |
| DDT | 0.002 | 1 | 250-300 | 0.05-0.2 |
| Dieldrin | 0.1 | 0.25 | 40-87 | 0.05 |
| Dimethoate (Perfekthion) | 25000 | | 250 | |
| Dinitro-o-cresol | 130 | | 30 | 6-13 |
| Endosulfan (Thiodan) | | 1 | 40-50 | 0.001-0.01 |
| Endrin | Practically insol. | 0.25 | 5-45 | 0.001-0.008 |
| Formaldehyde | Readily soluble | | 2400 | 150 |
| Lindane | 10 | 0.5 | 88-125 | 0.2-0.3 |
| Malathion | 145 | 15 | 1375-2800 | 0.1-1 |
| MCPA | Salts readily soluble | | 700 | 35 |
| Meta-Systox | 3300 | | 40-60 | 4-7.5; 100 |
| Methoxychlor | Practically insol. | | 5000-7000 | 0.05 |
| Paraquat | Readily soluble | | 150 | 32 |
| Polyram | Practically insol. | | 10000 | 32 |
| Parathion (E 605) | 24 | | 6-15 | 3 |
| Toxaphene | 3 | 0.1 | 40-120 | 0.006-0.2 |
| 2,4-D | Salts readily soluble | | 375 | 5-75-1160 |

effects. These were compiled from the reports of Perkow[246], Beran[24], An Der Lan[10], Lüdemann and Neumann[207] as well as Meinck-Stoof-Kohlschütter[218].

Many pesticides have a very intense odor and are thus detectable in water even in low concentrations. Quentin and Huschen-beth[254] report the following threshold odor concentrations in $\mu/l$:

| Aldrin | 1.3 | DDT | 160 | Malathion | 4 |
|---|---|---|---|---|---|
| Heptachlor | 9 | Lindane | 1000 | Parathion | 30 |
| Dieldrin | 27 | Toxaphen | 5 | Dimethoat | 50 |
| Methoxychlor | 100 | | | | |

Other data on threshold odor numbers, characterization of the odor and toxicities of 32 weed killers and pesticides can be found in Sigworth[289]. Naturally the odor also depends on the purity of the industrial product.

## 14.2    Sampling, Enrichment, Isolation

*Sampling.* The volume of water sample required for the analysis differs from one case to another and may range from a few ml (for example, for gas chromatography) up to about 50 l if special enrichment procedures are necessary.

Weil and Quentin[325] as well as Quentin and Huschenbeth[254] have pointed out that adsorption processes may lead to great losses of active ingredients when water samples are stored in plastic containers. The contents of glass vessels also need to be worked up in their entirety and the substance adsorbed on glass walls must be recovered by extracting with petroleum ether for 1 h.

On the other hand, Weil and Quentin[326] have described selective sampling by suspending polyethylene films (15 x 10 cm, 20-25 $\mu$m thickness) in a body of water. After 3 days the film is extracted with 1 ml petroleum ether in the absence of air and the extract is subjected to gas chromatography.

*Isolation.* The pesticides are frequently isolated from a water sample in a manner similar to the hydrocarbons. Analytical methods of excellent sensitivity exist here, too, which may be carried out directly on the water sample; as a rule, however, enrichment or extraction from a larger volume of water will be necessary.

If fairly large amounts of pesticides are to be isolated so that large volumes of water need to be treated, adsorption on activated carbon as described on p. 149 for hydrocarbon isolation can be recommended. It is also possible to work up the entire CCE-extract for the desired pesticides.

While the extensive analogy of physical properties of hydrocarbons permits their combined isolation, this often is not the case for pesticides. Many are almost insoluble in water and can be extracted completely by extracting once with $CCl_4$, chloroform or petroleum ether, for example.

The period of extraction differs according to literature data. They range from a few minutes to several hours. Weil and Quentin[327] extract 1.8-1.9 l of water sample with 20 ml petroleum ether for 12 h in 2 l rectangular bottles with ground stoppers. After settling for 1 h, and attachment of a microseparator (Fig. 3.2a), the petroleum ether layer is collected in the narrow section of the right riser tube by pouring water into the funnel tube on the left and an aliquot (for example, 15 min) is withdrawn with a pipette. This method permitted the recovery of 98-105% of 10 $\mu$g of added halogenated hydrocarbon insecticides.

The choice of extractant is based not only on solubilities but also on the requirement of later procedures. If the analysis of halogenated hydrocarbons is based on the presence of the halogen, as is the case in TLC identifications by the separation of silver, and particularly GC with the use of an electron capture detector or with automatic coulometric titrations, highly volatile hydrocarbons, for example, petroleum ether or isooctane, are necessary. For an IR-spectrophotometric reading, carbon tetrachloride is selected which is transparent in the important spectral regions.

Ether or petroleum ether often serves as the extractant of water-soluble pesticides.

The Kuderna-Danish apparatus (see Gunther and Blinn[119]) (Fig. 3.2b) is often recommended for a rapid and loss-free evaporation of highly dilute extract sloutions. Depending on the boiling point and quantitative ratio, different parts of the apparatus can be used. For example, 300 ml petroleum ether can be concentrated to 2 ml in 10 min.

Bevenue, Kelley and Hylin[27] have described the necessary measures for ultrapurification of organic solvents and laboratory apparatus for gas-chromatographic analyses in the ppb-range.

## 14.3 Separation and Identification

Thin-layer chromatography and gas chromatography, which have very high separating power, are usually used to separate similar pesticides. The latter procedure has the advantage that a differentiation of different groups of compounds is possible by the use of specific detectors which may also be used simultaneously (see 6.4.2). While a flame ionization detector indicates all organic compounds with CH-bonds, the electron capture detector responds preferentially to halogenated hydrocarbons, and the alkali ionization detector to phosphorus-containing compounds. Coulometric detectors can be adjusted selectively to certain titration methods.

Most recently column-chromatography techniques have been refined to such a degree in apparatus that their sensitivity and separation power approach those of gas chromatography (see 6.3).

A very specific detection of individual insecticides is possible on the basis of IR-spectra; however, these techniques require somewhat larger quantities of sample than the chemical methods.

It is probable that chlorinated hydrocarbons continue to predominate quantitatively as pesticides, even though there are serious objections against their use because of their stability and possibility of enrichment in animal organisms (for example, edible fish) and human organs; these objections have already led to their prohibition in many places.

### 14.4    Determination of Chlorinated Hydrocarbons on the Basis of the Chlorine Content; Chlorine Cleavage with Metallic Sodium

According to Koppe and Rautenberg[175] 1 l of water sample is extracted with 20 ml isooctane (2,2,4-trimethylpentane, Uvasol 4718, Merck) for 3 h. Ten ml of the extract are mixed with 10 ml isobutanol, 0.5 g of metallic sodium, freed from scale and divided into 6-8 pieces, is added and the mixture is boiled for 2 h with refluxing up to complete dissolution of the sodium. Five ml water are added and the solution is acidified with 8 ml 20% nitric acid. The aqueous phase is separated in a separatory funnel and the non-aqueous layer is washed twice with 6 ml portions of water. The aqueous solutions are collected in a 50 ml test tube with a mark at 30 ml and diluted with water to the mark. One ml of a 2% gum arabic solution in 0.4% nitric acid and 0.5 ml of 0.2 N $AgNO_3$ solution are added and the solution is irradiated for 15 min with a UV lamp at a distance of 20 cm. The absorbance is read at 600 nm (filter 618) against a solution of 1.8 g $NaNO_3$ + 2 ml $HNO_3$ + 1 ml gum arabic + 0.5 ml $AgNO_3$ in 30 ml. A calibration curve is constructed with test solutions, for example, from hexachlorocyclohexane, in quantities corresponding to 50-500 $\mu$g Cl.

## 14.5 Colorimetric Methods of Determination

Colorimetry and spectrophotometry primarily are tools of older methods of determination. Because the reaction steps are frequently complex, a considerable amount of sample is required and the specificity is not very good, they are used only rarely today.

For example, lindane γ-hexachlorocyclohexane,, benzene hexachloride) is determined in surface water according to Hancock and Laws[125] by adsorption on activated carbon, elution by boiling with a mixture of glacial acetic acid and acetic anhydride, dehalogenation of the acetic-acid extract with zinc into benzene, nitrogenation of the latter into dinitrobenzene, and finally, colorimetry according to Schechter and Hornstein[273].

DDT is determined in river water according to Berck[25] by extraction with ether/n-hexane (3 + 1), evaporation and nitrogenation of the residue into the tetranitro-compound according to Schechter and Haller[272], followed by colorimetry.

Parathion (E 605) is reduced with zinc and hydrochloric acid with the formation of an amino-group from the nitro-group. The reduction product is diazotized with sodium nitrite in the customary manner and coupled with N-(1-naphthyl)ethylenediamine into an intensely red dye which is then colorimetered.

## 14.6 Analysis of Halogenated Hydrocarbon Insecticides by IR-Spectrophotometry

According to Rosen and Middleton[261], a fairly large water sample is filtered on activated carbon, the latter is extracted with chloroform (see CCE method, p. 149), the chloroform is evaporated, the residue is again dissolved in a small amount of chloroform, filtered through an activated alumina column and eluted with three times the volume of chloroform (referred to the column volume), during which more strongly adsorbed components, particularly oxygen-containing compounds, are retained on the column. The eluate is evaporated, the residue is taken up with 0.2 ml carbon disulfide  and read in an IR-spectrophotometer between 5 and 15 $\mu$m. The IR-spectra of 7 chlorinated hydrocarbon insecticides are shown in Fig. 14.6.

Other examples on the IR-spectrophotometry of pesticides are given by Blinn and Gunther[30] as well as by Blinn[29], and by Crosby and Laws[61].

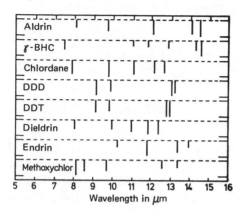

Fig. 14.6—Schematic diagram of the IR-spectra of 8 commercial insecticides.

## 14.7     Application of High-Efficiency Liquid-Liquid Chromatography

Waters, Little and Horgan[323] and a diagram from the Waters Company (Fig. 14.7) describe the separation of 12 commercial insecticides by high-efficiency liquid-liquid partition (see 6.3).

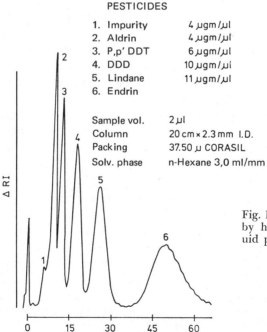

PESTICIDES

| | | |
|---|---|---|
| 1. | Impurity | 4 µgm/µl |
| 2. | Aldrin | 4 µgm/µl |
| 3. | P,p' DDT | 6 µgm/µl |
| 4. | DDD | 10 µgm/µl |
| 5. | Lindane | 11 µgm/µl |
| 6. | Endrin | |

| | |
|---|---|
| Sample vol. | 2 µl |
| Column | 20 cm × 2.3 mm I.D. |
| Packing | 37.50 µ CORASIL |
| Solv. phase | n-Hexane 3,0 ml/mm |

Fig. 14.7.—Pesticide separation by high efficiency liquid/liquid partition.

It was carried out with 2 μl of a mixture of (2) 4 μg Aldrin/μl, (3) 4 μg DDT/μl, (4) 6 μg DDD/μl, (5) 10 μg Lindane/μl and (6) 11 μg Endrin/μl. The column (20 cm long, 2.3 mm I.D.) was packed with 37.5 μm Corasil. The solvent was n-hexane (3 ml/mm) and the refractive index served for identification.

## 14.8 Thin-Layer Chromatographic Separating Methods

Older techniques of paper chromatography (see, for example, Evans[78]) for chlorinated hydrocarbon insecticides on paraffin-impregnated paper with acetone or for organophosphorus pesticides according to Getz[103] have been replaced by the techniques of thin-layer chromatography.

Bäumler and Rippstein[23] separate chlorinated hydrocarbon insecticides on activated $Al_2O_3$ plates with hexane. The following Rf-values were obtained with a development time of 45 min:

| | Aldrin | DDT | Hexachlor-cyclohexane | Dieldrin | Methoxy-chlor |
|---|---|---|---|---|---|
| $R_F$ | 0.78-82 | 0.59-62 | 0.39-41 | 0.17-19 | 0.10-12 |

Detection of the separated spots was carried out by spraying with 0.5% solution of N,N-dimethyl-p-phenylenediamine hydrochloride in sodium ethoxide solution, wetting with water and identification of the violet-greenish spots under UV-irradiation. Dyatovitskaya and Gladenko[70] spray the plates with ammoniacal $AgNO_3$ solution and irradiate with UV-light; Petrowitz[247] sprays the plates with monoethanolamine, heats them for 20 min to 100°, sprays with $AgNO_3$ solution in nitric acid and irradiates with UV-light. This permits the detection of 1-2.5 μg of insecticide.

Taylor[304] uses silica gel plates and $CCl_4$ solvent for chlorinated hydrocarbon insecticides. Detection is carried out by spraying with monoethanolamine + $AgNO_3$ and UV-irradiation.

Hamilton and Simpson[124] use Florisil plates for the same purpose. They report the Rf-values of 90 compounds.

Silica gel G is used for the separation of phosphorus insecticides. Bäumler and Rippstein[23] use hexane-acetone (4 + 1) as a solvent. The spots are detected with palladium(II) chloride in a weak hydrochloric acid solution. Fischer and Klingelhöller[85] separate thiophosphoric acid esters with methylene-methanol-$NH_3$ and perform the detection with iodine azide solution with the formation of white spots. The detection sensitivity amounts to 1-5 $\mu$g.

Wildbrett, Ganz and Kiermeier[337] described other separating procedures; Goodwin, Goulden and Reynolds[111] report on phosphorus insecticides. Abbott, Egan and Thomson[3a] cite Rf-values of 16 chlorinated hydrocarbon insecticides for 10 separating systems (silica gel, $Al_2O_3$ with cyclohexane-paraffin oil-dioxane as well as n-hexane as solvent systems).

*Microcircular thin-layer technique according to Koppe and Rautenberg[175].* As described in 14.4, a water sample of 1 l is extracted with 20 ml isooctane. The extract is concentrated to a known quality and 0.01 ml is applied on a thin-layer plate with a pipette in such a way that a circular spot of 6-8 mm diameter results. The thin layer is prepared from 3.5 g silica gel G (Merck, 8129) and 15 g aluminum oxide G (fine-grain, neutral; Merck, No. 1090) in 40 ml water in a layer thickness of 0.15 mm with a surface density of 5 mg/cm². The plates are dried for 30 min at 110° and cleaned by diffusion with alcohol before use.

As standard solutions, 0.01 ml portions of hexachlorobenzene in isooctane solution containing 0.1-0.5 $\mu$g organically bound chlorine are applied beside the sample solution.

The solvent is evaporated in an air stream and sufficient ethanol (about 0.001-0.002 ml) is applied on the sample residue that the area of the circle becomes of the same size as it was with isooctane. After evaporation of the alcohol, the spot is sprayed with spray solution I, irradiated for 4 h with a UV-lamp at 50 cm distance, and the gray circular rings are compared with those of the test solutions. Spray solution I: 34 ml water + 10 ml N/5 $AgNO_3$ + 2 ml nonylphenolethoxylate solution (10% Arkopal solution of Henkel-Dehydag in water) + 4 ml concentrated ammonia.

Easily cleavable chlorine can be determined by spraying with a mixture of equal volumes of 10% aqueous $NH_3$ solution and 0.1 N $AgNO_3$ solution, followed by UV-irradiation for 2 h. Chloroparaffin serves as the reference substance.

*Determination of organochlorine herbicides by thin-layer chromatography.* Abbott, Egan, Hammond and Thomson[4] describe the separation of MCPA, MCPB, 2,4-D, 2,4-DB, 2,4,5-T and Dalapon.

Extraction: A water sample of 100 ml is acidified with 5 ml 6 N $H_2SO_4$ and, after standing for 5 min, is extracted twice with

100 ml portions of ether. The ether solutions are extracted twice with a mixture of 150 ml 8% $Na_2SO_4$ solution and 5 ml 2.5 N NaOH, and are then discarded. The alkaline-aqueous extracts are acidified with 10 ml 6 N $H_2SO_4$ and extracted with 150 and 100 ml of ether. The ether solutions are dried and evaporated on $Na_2SO_4$, the ether residues are blown out and the residue is dissolved in 40 µl ethylacetate.

TLC treatment: A mixture (30 g) of 60 parts kieselguhr G and 40 parts silica gel G is suspended in 60 ml water and used to coat 4 plates (20 x 20 cm) in 250 µm thickness. Activation is carried out at 120° for 2 h. A mixture of 10 ml paraffin oil, 30 ml benzene, 20 ml glacial acetic acid and 200 ml cyclohexane serves as the solvent system. The development distance amounts to 15 cm. After brief drying in air, the plates are treated for 10 min at 120° in a drying oven, sprayed with 0.5% alcoholic $AgNO_3$ solution and again conditioned at 120° for 10 min. They are then irradiated with UV-light for 10 min. The spots are compared visually or by densitometry with test spots of 2-10 µg of the above-mentioned standard herbicides with regard to their Rf-value and intensity. The Rf-values increase in the order of Dalapon 2,4-D-2,4,5-T — MCPA — 2,4-DB — MCPB. The recovery of 0.1-25 mg herbicide per 100 ml water amounted to 80-100%.

*TLC determination of triazine herbicides according to Abbott, Bunting and Thomson*[3]. A water sample of 200 ml is adjusted to pH 9 with $NH_3$ and extracted twice with 25 ml portions of dichloromethane. The $CH_2Cl_2$ solution is dried on $Na_2SO_4$ and evaporated to dryness in the Kuderna-Danish evaporator (see Fig. 3.2b). The residue is dissolved in 50 µl hexane and developed on silica gel G plates with chloroform-acetone (9 + 1) for 35 min. The plates are allowed to dry and are sprayed with 0.5% acetone solution of brilliant green, followed immediately by exposure to bromine vapors. The spots formed are marked and the area is measured. The square root of the area is plotted against the logarithm of the weight of test substance utilized in a calibration line. The limit of detection amounts to 0.5-1 µg.

*TLC and GC determination of chlorinated hydrocarbon pesticides in carbon-chloroform extract.* Smith and Eichelberger[291] determine chlorinated hydrocarbon pesticides in the CC-extract (see p. 149) by prepurification on a thin-layer plate and subsequent gas chromatography.

Thin layer: Silica gel G, 0.25 mm thickness, dried at 130° for

30 min. Solvent: $CCl_4$; development distance 10 cm. The samples are applied in 2 series, one series serving for localization. Color reagent: spraying with a solution of 1.7 g $AgNO_3$ in 5 ml water, addition of 10 ml 2-phenoxyethanol, diluted to 200 ml with acetone. The plates are then exposed to UV-radiation for 4-7 min. A different spray which does not destroy the substances is a solution of 10 mg rhodanine B in 100 ml ethanol. Under UV-irradiation the pesticide spots appear dark on a fluorescent background. The spots are marked, scraped off, the powder is aspirated into a small eye dropper with a glass wool plug; it is extracted from the latter with a small amount of ether-petroleum ether $(1 + 1)$, the solution is evaporated to a small volume and 5 $\mu$l of this are injected into the gas chromatograph.

The column (900 x 3 mm) is packed with Dow silicone grease on Chromosorb W; carrier gas 5% methane in argon, 50 ml/min; electron capture detector; column temperature 175°.

## 14.9    Gas-Chromatographic Methods of Pesticide Analysis

As mentioned in 6.4.2, gas-chromatographic analysis can be adapted to the expected pesticide group by the selection of specific detectors.

A determination of traces of lindane in ground water has been described by Ströhl[300]. The water sample is extracted with ether/hexane $(3 + 1)$ according to Reploh and Buenemann[258] and is concentrated. One $\mu$l of solution is injected into a 4 m column (2% silicone elastometer SE 30 on Diatoport 60-80 mesh). The carrier gas consists of 5% methane in argon at 80 ml/min. The temperatures amount to 175° in the column and 200° in the electron capture detector.

For a lindane determination in water Young[342] uses 1 m glass columns packed with 5% QFI (PE) on Chromosorb W. The detection sensitivity with the electron capture detector falls into the nanogram-picogram range (Drescher[69]).

Herzel[133] also describes the application of a thermionic detector for organic phosphorus compounds as well as of a microcoulemetric detector for compounds containing halogens, sulfur and phosphorus.

Pionke, Konrad, Chesters and Armstrong[249] (see also Konrad, Pionke and Chesters[172]) extract a 500 ml water sample with 25 ml benzene for 2 min. The benzene solution is concentrated to 1 ml in an air stream. One $\mu$l of this solution is injected into a 2 m column packed with 10% DC 200 on 60-80 mesh Gas-Chrom O for analysis of chlorinated hydrocarbons. Injection temperature 235°, column temperature 195°, electron capture detector temperature 200°, voltage 50 V, carrier gas $N_2$, 125 ml/min.

For the determination of Parathion, gas chromatography is performed under the following conditions: column of 2 m length packed with 10% DC 200 on 80-90 mesh Chromosorb W; column temperature 200°, injection and outlet temperature 240°; KCl-thermionic detector at 200°; 300 V voltage.

Weil and Quentin[325] perform their model experiments with halogenated hydrocarbons by injecting the petroleum ether extract (see 14.2) into a 2 M glass column packed with 1% silicone elastomer SE 30 on Chromosorb G, treated with AW-DMCS. Temperatures: column 195°, injection block 250°; purified nitrogen at 40-75 ml/min as carrier gas; electron capture detector with tritium, 104 V, 195°. Retention data (aldrin $=$ 1) lindane 0.87, heptachloroepoxide 1.60, dieldrin 2.30, DDT 4.00, methoxychlor 7.80.

Faubert, Maunder, Egan and Roburn[83] describe the calibration and evaluation of the electron capture detector readings with lindane.

Drescher[69] also described the determination of chlorine insecticides in the picogram range with an electron capture detector. The glass separating column (1.5 m, 2.5 mm) is packed with 5% DC 11 on Chromosorb W/AW/DMCS; 200°; $N_2$ 45 ml/min.

Kneip, Beasley, King and Dean[167] described the separation and determination of lindane, heptachlor, aldrin, heptachloroepoxide, endrin, DDT and methoxychlor in dual columns (180 cm x 5 mm, 10% Dow Corning 200 on 100-120 mesh Gas-Chrom Q [Applied Science Lab]) each with one electron capture detector; the second column serves as a blank reference system. Temperature 200°, preferably programmed at 3°/min from 200 to 260°. The nitrogen was purified on titanium sponge at 400° and molecular sieve 5 A. Detector voltage 22.5 V. Up to 0.1 ml petroleum ether solution can be injected. Samples of 0.5-2.5 ng halogenated hydrocarbons were dissolved in 250 ml petroleum ether, evaporated to 5 ml and 0.1 ml portions were subjected to gas chromatography. A relative standard deviation of $\pm$ 26% was found.

Kawahara, Moore and Gormann[161] reported on a combination of gas and thin-layer chromatography for the determination of chlorinated hydrocarbon insecticides.

Kawahara, Lichtenberg and Eichelberger[160] also described the determination of parathion and methylparathion by thin-layer chromatography in the presence of chlorinated hydrocarbons.

Wilder[338] recommends that for a gas-chromatographic analysis of 2,4-D salts with the flame ionization detector, the extract be esterified with diazomethane.

*Gas-chromatographic technique of the APHA Standard Methods, 13th Edition (1971).* The 4 l water sample is extracted twice with 10 ml portions of hexane for 15 min. The hexane extract is dried on $Na_2SO_4$, evaporated to 7 ml in the Kuderna-Danish evaporator and diluted with hexane in a 10 ml graduated flask. If a preliminary gas-chromatographic run with 5 μl shows that a higher concentration or additional purification is necessary, the hexane solution is loaded on a 400 x 20 mm Florisil column. The chlorinated hydrocarbons are eluted from this with 200 ml of a mixture of 1 vol. diethylether and 3 vol. petroleum ether.

Two 1.80 m columns are described: a nonpolar packing of 10% DC 200-silicone oil on 80-90 mesh Anakrom ABS (Analab Inc., Hamden, Conn.), and a "polar" packing consisting of 4% SE 30 (or DC 200) + 6% QF 1 (fluorosilicon) on the same support. Electron capture detector; temperatures 230° in the injector, 180° in the heater, 200° in the detector. Carrier gas $N_2$, 120 ml/min.

Further data on experimental and instrumental details, calibration, attainable accuracy and the relative retention times of 47 pesticides are presented in the original paper.

*Organophosphorus insecticides.* Although organophosphorus insecticides are not as stable as chlorinated hydrocarbons, they remain in water without decomposition for a sufficient period of time that it may become necessary to analyze them, for example, in a surface water. Older methods are based on color reactions; more recently, GC or TLC separating techniques have been employed. Isolation from water samples by extraction with organic solvents is more difficult than with chlorinated hydrocarbons because these compounds generally have a higher water solubility.

*Chromatographic techniques according to Askew, Ruzicka and Wheals*[13]. After dissolving 20 g $Na_2SO_4$ in 1 l of water sample, the latter is extracted three times with 50 ml portions of chloroform. The chloroform solutions are dried on anhydrous $Na_2SO_4$ and the chloroform is evaporated.

Three column packings are described for the gas-chromatographic determination:

I. 2% Apiezon L and 0.2% Epikote 1001

II. 4% SE 30 and 0.4% Epikote 1001

III. 2% XE 60 and 0.2% Epikote 1001

The columns have a length of 150 cm, inside diameter of 3 mm; the support is Chromosorb G, WA, DCS-treated. A phosphorus-specific alkali ionization detector is used (see 6.4.2).

The TLC separation is carried out on 250 $\mu$m thick silica gel plates activated at 120° for 2 h. Hexane-acetone (5 + 1) is a generally applicable solvent system. For detection the plates are first sprayed with hydrochloric acid, the layer is then covered with a glass slide and heated in an oven to 180° for 30 min in order to decompose the organophosphorus compounds; it is then cooled and sprayed with ammonium molybdate solution, heated again for 5 min in the oven, and after cooling, is sprayed with $SnCl_2$ solution. Background of the blue spots can be bleached by brief exposure to ammonia vapors.

In a table the authors list the gas-chromatographic retention times as well as the Rf-values of the TLC separation for 40 commercial organophosphorus insecticides analyzed with the above-described columns. A separation by gel chromatography on Sephadex LH 20 columns is also described.

The limits of detection with thin-layer chromatography amount to 1 $\mu$g or 1 ppb per liter of water sample. GC is more sensitive by one order of magnitude.

Special data on the gas-chromatographic determination of 20 organophopsphorus insecticides (with no reference to water analysis) can be found in the study of Egan, Hammond and Thompson[72].

Cohen, Norcup and Ruzicka[53] describe the determination of carbamate insecticides by GC in the form of 2,4-dinitrophenol derivatives with the electron capture detector; El-Dib[75] reports on their TLC.

### 14.10   Polychlorinated Biphenyls

Recently polycholrinated biphenyls are frequently used in industry as heat transfer agents (for heating and cooling) as well as plasticizers and insulating materials. They have repeatedly been introduced into surface water by industrial effluents and have been detected there. Their toxicity is lower than that of chlorinated hydrocarbon insecticides, but their chemical stability has already led to enrichment in animal (seagulls) and human organs via various food chains, so that their determination is becoming of increasing interest in water analysis.

The polychlorinated biphenyls (for example, Chlophen A 50) involve mixtures of different degrees of chlorination. Gas chromatography therefore usually results in a fairly large number of peaks. The separation of insecticidal chlorinated hydrocarbons presents some difficulties and they therefore often interfere with the analysis.

This has recently been discussed by Bauer[20a] and Herzel[133a] as well as Bevenue and Ogada[27a].

Veith and Lee[317a] have mentioned the uncertainties in the identification of polychlorobiphenyls in gas-chromatographic peaks which cannot be assigned to the known insecticides, so that their detection is frequently uncertain. Further studies are necessary to clarify this problem before promising control measures can be taken.

*Chlorinated hydrocarbon insecticides in the presence of polychlorobiphenyls.* Ahling and Jensen[6] isolate the mentioned substances by liquid-liquid partition on a lipophilic support. A 30 cm column is packed with Chromosorb W (HMDS-treated) impregnated with 10% Carbowax 4000-monostearate + 30% n-undecane (in acetone solution). The water sample is transported through the column at a rate of about 100 ml/min.

It is also possible to dry an entrainment-precipitate with $Al_2(SO_4)_3$ and extract it subsequently with acetone, followed by a petroleum ether extraction.

The still wet column is loaded with petroleum ether. The first 10 ml are collected, shaken with an equal volume of concentrated sulfuric acid, centrifuged, cooled on dry ice, the petroleum ether layer is decanted and 10 $\mu$l of this are analyzed in the gas chromatograph.

The glass GC column of 160 cm length and 2 mm inside diameter is packed with 80-100 mesh Chromosorb W (treated with HMDS) containing either 4% methylsilicone oil SF 96 or 8% fluorosilicone oil QF 1. Electron capture detector (Varian Aerograph 204); purified nitrogen at 30 ml/min as carrier gas. Temperatures 205° on the detector, 220° on the injector and 190° in the column. Retention time of 20 min for DDT and 10-40 min for polychlorodiphenylene. Pesticide yield: 50-100%, PCB yield 93-100%.

It is possible to detect 10 ng/m³ lindane in 200 l water (200 l sample volume).

## 14.11 Gas-Chromatographic Determination of Dichloropropionic Acid

Dichloropropionic acid (Dalapon) is used as a complete herbicide. The gas-chromatographic determination can be carried out by the method of Frank and Demint[92]:

The water sample is almost saturated with NaCl (34 g per 100 ml) and acidified to pH 1 with HCl. A volume of 500 ml is extracted consecutively with 50, 25 and 25 ml ether. The ether extracts are extracted three times with 20, 10 and 10 ml of a NaCl-saturated 0.1 N NaHCO₃ solution adjusted to pH = 8 with NaOH. The aqueous extracts are adjusted to pH = 1 with HCl and extracted with 5, 4 and 4 ml ether. The ether is concentrated to 2 ml in an air stream at room temperature and an equal volume of 0.5% diazomethane solution in ether is added. After standing for 10 min, gas chromatography is performed.

The authors use the Aerograph (Model 204) with electron capture detector, steel column of 1.5 m length and 3 mm I.D. packed with 60-80 mesh Chromosorb P, HMDS-treated, with 10% FFAP (esterification product of Carbowax 20 M and 2-nitroterephthalic acid). Temperatures: column 140°, detector 195°, injection block 215°. Carrier gas N₂ purified with molecular sieve, 30 ml/min. Injection volume 1-5 µl.

Calibration is performed with 2,2-dichloropropionic acid as the test substance.

*GC of pentachlorophenol.* Pentachlorophenol is frequently used as a wood preservative and pest control. Rudling[265] uses the following procedure for its determination in water:

A water sample of 100 ml is acidified with 2 ml concentrated H₂SO₄ and extracted for 1 min with 100 ml n-hexane. The hexane phase is extracted twice with 2 ml portions of 0.1 M borax solution; the aqueous borax extracts are shaken for 1 min with 0.5 ml n-hexane and 4 µl of an acetylation mixture (2 ml pyridine + 0.8 ml acetic anhydride). The hexane phase is used for GC.

Column of 1 m x 1.5 mm I.D. packed with 5% QF 1 on Varaport 30, 100-120 mesh; carrier gas N₂ 25 ml/min; electron capture detector. Temperatures: 160° in the injector, 150° in the column and 170° in the detector. Pentachlorophenol has a retention time of 3 min; the limit of detection is 0.1 ppb.

## 14.12   Group Identification on the Basis of Cholinesterase Inhibition

The toxic effect of some groups of pesticides, particularly the phosphate and carbamate insecticides, on pests is based on inhibition of the activity of cholinesterases, which are important for life. The presence of such agents in water samples thus can be detected and quantified by suitable enzyme activity determinations. Giang and Hall[104] as well as Schutzmann and Barthel[280], and Weiss and Gakstetter[329] make use of cholinesterases for this purpose; the Technicon Autoanalyzer can serve the same purpose (Winter and Ferrari[340]).

Weil and Quentin[328] employ the pseudocholinesterase inhibition for the purpose. The enzyme is available commercially (Schuchardt, Catalog No. 364), and so are the substrate and indicator (Merckotestcholinesterase, Merck, Catalog No. 3337).

The pesticides are extracted conventionally with ethylether or benzene, 0.5 ml 2-(2-ethoxyethoxy) ethanol solution is added and the extracts are concentrated, for example, in the Kuderna-Danish evaporator; the residue is treated with 2 ml Chromogen solution (25°) and 100 $\mu$l pseudocholinesterase solution and after 30 min, the substrate solution is added. The mixture is shaken and after 30-60 sec is read in a spectrophotometer with a 1 cm cuvette to determine the time required for a 0.100 increase of the 412 nm absorbance. The reagent blank value is determined at the same time.

The per cent inhibition is calculated on the basis of the fact that the time required for a 0.1 change of the absorbance in seconds is in inverse proportion to the active enzyme concentration present.

The concentration of the enzyme stock solution should amount to 0.5 International Units (U) $\pm$ 10% (1 U refers to the quantity of enzyme which converts 1 $\mu$mol of substrate in 1 minute). It is obtained according to the formula:

$$\frac{9.25}{t} U/ml \ (25°),$$

where t is the time in seconds required for a 0.100 absorbance change.

The inhibition in per cent is obtained by the following formula:

$$\% \ \text{inhibition} = (1 - \frac{t_1}{t_2}) \cdot 100$$

$t_1$ = time for a 0.1 absorbance change in the blank value in sec,

$t_2$ = time for a 0.1·absorbance change in the presence of the sample extract in sec.

The pesticide concentration is expressed in "parathion equivalents." This is the volume of parathion which produces the same inhibiting effect in an extract from 1 l of water. It is determined on the basis of a calibration curve with water of known parathion concentration.

### Reagents

Chromogen solution: 0.25 mmol 5,5′-dithio-bis-2-nitrobenzoic acid in 0.02 M phosphate buffer, pH = 7.7; 0.0272 g monobasic potassium phosphate and 0.3304 g dibasic sodium phosphate dihydrate as well as 0.01 g 5,5′-dithio-bis-2-nitrobenzoic acid are dissolved in 100 ml water.

Substrate solution: 0.12 M S-butyryl thiocholine iodide; 3.806 g S-butyryl thiocholine iodide are dissolved in 100 ml water.

2-(2-ethoxyethoxy) ethanol solution: 2% solution in diethylether.

Pseudocholinesterase solution: 1 mg enzyme (for example, No. CH364 of Schuchardt Co.) is dissolved in 10 ml physiological saline (0.9 g NaCl in 100 ml distilled water). The activity of this solution is determined according to the above instructions and adjusted to 0.5 International Unit.

The method was employed by the authors for 17 phosphorus pesticides and 4 carbamates available on the market. Inhibitions of 5% or pesticide quantities of less than 1 $\mu$g can still be detected.

# 15. UREA

Urea is not only an end product of animal and human metabolism (man excretes an average of 30 g urea/day), but is also a product of the chemical industry. It is used in agriculture as a fertilizer and in industry for the production of synthetic resins. Its application on roads and bridges in winter as a spreading salt instead of rock salt is under consideration for locations which are subject to corrosion.

Thus, the presence of urea in water need not necessarily be considered proof of the discharge of residential waste effluents. Depending on the presence of enzymes (ureases) it has a limited life which amounts only to a few hours in residential effluents.

*Urea determination with NaOCl and phenol.* While the well-known indophenol reaction of ammonia with NaOCl and phenol leads to a blue product with an absorption maximum at 630 nm, the same reaction with urea results in a yellow color with its absorption maximum at 454 nm.

*Urea determination method according to Emmet*[76]. A water sample of 20 ml in a 25 ml graduated flask is treated with consecutive additions of 3.0 ml distilled water, 0.30 ml NaOCl reagent, 0.50 ml NaOBr reagent (with 1-2 seconds of stirring after each addition) and 0.30 ml hydrochloric acid as well as the necessary quantity of hydrochloric acid to adjust the 0.50 ml NaOBr solution to pH 7.5, all with vigorous stirring with a magnetic stirrer; after stirring for 10-20 sec, 0.50 ml borate buffer is added, followed immediately by 0.30 ml phenol reagent. After stirring for 10-20 sec, the solution is allowed to stand for 10-20 min and the absorbance is then read at 454 nm.

A mixture of water sample and reagents without phenol serves to determine the blank value. The limit of detection for urea amounts to 5 $\mu$g/l.

*Reagents*
2.5% NaOCl solution.
0.2% NaOBr from 2.0 g NaOH and 0.75 ml bromine.

HCl: 0.25 N with an addition of 0.3% $MgCl_2 \cdot 6\ H_2O$.

Borate buffer: 0.6% $H_3BO_3$, 0.3% KCl, 0.2% $H_2O_2$, 0.25% NaOH, pH = 9.6.

Phenol solution: 1.0 M in 95% ethanol.

*Urea determination with xanthydrol in swimming pool water.*
According to Fuchs[97] 500 ml—5 l of water sample are treated with an addition of 0.5 ml concentrated $HNO_3$ and 1 g sodium acetate, followed by evaporation to 10 ml. Acetic acid (15 ml) is added, followed by filtering and washing of the precipitate with 3 x 5 ml glacial acetic acid. The solution is precipitated with 5 ml xanthydrol solution (10% in methylalcohol) in 1 ml portions added every 10 min with swirling. The precipitate of dixanthylurea $C_{27}H_{20}O_3N_2$ is collected in a weighed porcelain filter dish, washed with methanol, dried breifly at 105° in a drying oven and weighed.

1 g dixanthylurea = 0.1428 g urea.

Brauss and Barth[36] have described urea determinations in swimming pool water based on an enzyme procedure with the use of the Biochemica test of Boehringer-Mannheim Corp.

# 16. URIC ACID

*Uric acid determination according to Kupchik and Edwards*[179]. Uric acid appears in the excrements of man and mammals, reptiles, birds and insects and is a relatively stable substance. It is a suitable indicator for the presence of residential pollutants. Its strong absorption at 292 nm is suitable for a spectrophotometric analysis; the attenuation of this absorption is read in a water sample after addition of uricase, the specific degradation enzyme. To allow application of this method for surface water, the reading must be taken at greater optical paths (50 cm). This is possible with the 50 cm cell of the Beckman spectrophotometer DU.

About 220 ml of the water sample, which had first been filtered on ordinary filter paper and then on a membrane filter, are adjusted to a pH of 9.3-9.4 by dropwise addition of N NaOH, and are then filled into the 50 cm cell. With a slit width of 1.0 mm the scale is adjusted to 20-80% transmittance at 292 nm and the reading is taken until a constant value has been reached. Subsequently a sufficient quantity of commercial uricase (depending on manufacturer) is added, the initial value is recorded and readings are taken every 5 min for 40 min until no further attentuation occurs (absorbance attenuation $A_a$).

The addition of the same volume of uricase is repeated and a possible increase S is recorded as the enzyme blank value. The total attenuation A is the sum of $A_a$ and S. Calibration with uric acid resulted in the equation $A = 0.0035 c - 0.0166$, where c = $\mu$g uric acid/l.

Even 5 $\mu$g uric acid per liter, corresponding to 1/100,000 of the concentration in human urine and 1/200 of the concentration in raw residential waste water can be detected. In residential waste water the parallelism between the uric acid content and the density of coliform bacteria represents a criterion for pollution.

# 17. COPROSTEROLS

According to Reichert[257] as well as Kussmaul and Muhle[179a] coprosterol is an indicator of the introduction of fecal matter into water. Nearly coprosterol-free water (less than 50 ng/l) proved to be free of fecal Coli bacteria, while their presence must be anticipated when the coprosterol concentrations are more than 100-200 ng/l.

Coprosterol is a reduction product of cholesterol formed by intestinal bacteria; man excretes 80-3500 mg/day or an average of 860 mg of coprosterol; the same is true for the most common domestic animals.

Reichert has modified a method of Murtaugh and Bunch[235] with the following instructions:

A water sample of 2 l is treated with 10 ml 20% sodium chloride solution and 5 ml concentrated hydrochloric acid; it is then shaken with 100 ml double-distilled n-hexane for 30 min. The n-hexane layer is extracted with 50 ml 70% ethanol and evaporated to dryness at 30° in a water-jet vacuum. The residue is saponified with 20 ml 7.5% KOH in 70% ethanol for 2 h with refluxing; the mixture is extracted with 20 ml water and 40 ml hexane, the hexane phase is extracted with 5 ml ethanol and concentrated to dryness at 50° in a nitrogen stream. The residue is taken up with 0.2 ml n-hexane and applied on a silica gel plate which has been activated at 110°. A sample of 4 $\mu$g coprosterol is applied on the left and on the right at the same time. Development is carried out for 45 min in chloroform/ether (9:1). After covering the development distance of the water sample, the test distances of the air-dried plate are sprayed with 10% alcoholic phosphomolybdate solution and placed into a drying oven at 110° for 5 min. The sample layer located between the test spots is scraped off, extracted with chloroform/ether (9:1) and evaporated in a $N_2$ flow; 50 $\mu$g bis (trimethylsilyl)acetamide are added to the dry residue which is then allowed to stand for 1 h in a closed container.

Of this solution 1 $\mu$l is further analyzed in a gas chromatograph equipped with a flame ionization detector. Column material:

Chromosorb W-HP coated with 3% SE 30 silicone elastomer. Temperatures: 260° in the column, 285° in the injector, 290° in the FID. Carrier gas $N_2$. A silylized coprosterol standard is loaded on to the column before and after the test run.

A clear signal is still produced by 1-2 ng/$\mu$l.

Kussmaul and Muhle[179a] use column chromatography instead of thin-layer chromatography for an improved separation of coprosterol and coprostanone—an oxidation product of coprosteraol—from cholesterol. The extract is saponified and evaporated as above and taken up in a small amount of hexane; it is then loaded on a column of 18 cm length and 6 mm I.D., packed with 3 g silica gel (100-200 $\mu$m particle size), and eluted with benzene (0.33 ml/min). Coprostanone is present in the 10-15 ml fraction, coprosterol in the 25-35 ml fraction and cholesterol in 40-50 ml.

Instead of silylation, Kussmaul and Muhle prepare the fluorine derivatives with trifluoroacetic anhydride (50 $\mu$l with respect to the anhydrous residue taken up in 100 $\mu$l hexane after evaporation of the benzene from the 10-40 ml fraction). After evaporation of excess trifluoroacetic anhydride at room temperature the residue is injected into the gas chromatograph. The glass column (1.7 m length, 3 mm I.D.) is packed with 10% OV-7 on Chromosorb W HP 80/100 (Varian). Column temperature 290°, FID and injector 320°; carrier gas $N_2$.

A sharper indication can be obtained in a mass spectrometer.

# 18. UROCHROMES

Urochromes are yellow pigments found in human and animal urine as the metabolism product of porphyrins; their determination in water is of interest in several respects. First of all, their presence is a clear indication of the introduction of fecal matter into drinking water and therefore is of greatest concern from the public health standpoint. In swimming pool water their concentration may be considered a criterion of the excreted urine.

Additional health concerns arise as a result of the studies of Hettche[134] who recognized that urochromes in water were a cause of endocrine goiter and developed an analytical method for their determination. A concentration of 2.5 mg/l in drinking water is the critical value for a strumigenic effect of urochromes (however, see also Wolter[341]).

According to Strackenbroch[299] urochromes in drinking water are decomposed by ozonation. Bucksteeg and Thiele[42] have reported on their behavior during biological waste water treatment.

The enrichment and isolation of urochromes from water samples according to Hettche is carried out by adsorption on aluminum oxide or entrainment precipitation with alum and ammonia. They are eluted with formic acid. Quantitative analysis is carried out by weighing or colorimetry. In both cases the presence of other pigments with similar analytical properties, such as humic acids and lignin derivatives, presents difficulties.

The elimination of humus as well as separation into several components of different effect are possible by paper chromatography.

*Adsorption on alumina according to Hettche.* A Schott glass frit 3 G 3 is sealed with a suspension of a few mg of Fuller's earth; aluminum oxide (for example, Riedel De Haen "for chromatography") suspended in water about 10 times and separated from suspended fines, is then applied in a layer of about 4-5 mm thickness. The number of drops should amount to about 2-3/sec with suction. The filter is washed with hot 85% formic acid until the filtrate is clear. Subsequently 125-500 ml of water sample (depending on the yellow color intensity) are aspirated through, followed by rinsing with 50 ml pure water. Elution is performed with 5 ml hot 85% formic acid, repeated 3-4 times, the optically transparent fractions are combined, their volume is measured and the absorbance is read in the 2 cm cuvette of the Elko II with filter S 38 (380 nm).

Calculation: With 20 ml eluate and 500 ml water sample, the absorbance value $E_{2cm}$ is multiplied by 18.4 to obtain the urochrome in mg/l. Example: 250 ml water sample produced 22 ml eluate and an absorbance of 0.500 in the 2 cm cuvette. This results

in $2 \times \dfrac{22}{20} \times 0.500 \times 18.4 = 20.2$ mg/l urochrome.

Determination by weighing: The eluate obtained with 85% formic acid is evaporated to dryness on a water bath, slurried in a small amount of 85% formic acid, separated from undissolved $Al_2O_3$ by centrifuging and the solution is again evaporated. If necessary, this procedure is repeated unitl a pure brown residue is obtained which is weighed. Subsequently, the material is ashed and the ash weight is subtracted.

*Isolation by entrainment precipitation and colorimetry.* A water sample of 500 ml in a 1 l graduated cylinder is treated with 20 ml 0.1 M alum solution and 1 drop phenolphthalein solution. Subsequently 2-4 ml 5% ammonia solution are added until a weak pink color can barely be recognized (pH = 7.6-7.8). The precipitate is allowed to stand for a few hours, the supernatant liquid is removed by siphoning and the balance is centrifuged. The residue in the centrifuge tube is dissolved in 5 ml 85% formic acid and if necessary, centrifuging is repeated. If iron is present, 0.5 ml 85% phosphoric acid is added. The mixture is transferred into a 50 ml graduated flask and diluted to the mark.

A visual color comparison can be made with a stock solution of 1.5 ml 1% potassium dichromate solution and 7.6 ml 5% cobalt nitrate solution, diluted to 100 ml. One ml of this stock solution per 50 ml water corresponds to 1 mg urochrome.

Photometry is carried out in the Elko II with the S 38 filter (380 nm). The absorbance with an optical path of 1 cm multiplied by the quantity of standard solution in ml (50 ml according to the above method) results in the "color index," and with the use of formic acid solutions, this value is multiplied by 1.9. One unit of this product corresponds to 0.5 mg urochrome. Example: a color index of 11.2 and (x 0.5 x 1.9) 10.6 mg urochrome in the formic acid solution correspond to an absorbance of 0.224 of 50 ml formic acid test solution in the 1 cm cuvette.

In the presence of humic acids an additional absorbance value is read with the S 53 filter (530 nm) according to Hettche. The difference in the logarithms of the two absorbance values, multiplied by 1000, furnishes a value of 0.90 for pure urochrome and 0.537 for humic acids (from brown coal).

According to Sattelmacher and Fürstenau[270a] an acceptable separation of humic acid and determination of the active urochrome B can be realized by esterification with diazo-octadecane and paper-chromatographic separation (see original study for method).

Strackenbrock[299] as well as Wolter[341] have described the chromatographic separation of urochrome A and B.

# 19. HUMIC ACIDS

As a group of compounds, humic acids are of interest in water chemistry since they involve materials of primarily nautral origin which are frequently present in surface and ground water in very different concentrations. They are the reason for the yellow color of such water when this is not produced by other pigments from residential and industrial waste effluents.

As such, humic acids are nontoxic and harmless from the standpoint of health; however, they considerably reduce the quality of drinking and industrial water in appearance and taste, and present difficulties in its use for cooking as well as beverage production as well as in many industrial applications, particularly in the preparation of boiler feed water with ion exchangers.

In the absence of other pigments of residential or industrial origin (urochromes, tempering pigments of the caramel type, ligninsulfonic acids and synthetic colorants), the color can serve as an identification and quantification of humic acids in the visible and UV region. However, very different color intensities are found in humic acids of different origin, as reported by Bohnsack[31].

Precipitation of humic acids from iron(III) chloride as well as lead nitrate have been described for their gravimetric analysis. After weighing of the corresponding humates, it is customary to determine the heavy metal content and subtract it from the product weight.

*Humic acid determination by UV absorption according to Fuchs and Kohler*[96]. The water sample is filtered and, if necessary, freed from heavy metals by treatment with a cation exchanger (for example, Lewatit S 100). The absorbance is read at 300 nm. The following absorbance values have been found in aqueous humic acid solutions (from brown coal) with a Zeiss Opton instrument in a 20 mm quartz cuvette:

Table 19.—Absorbance values of aqueous humic acid solutions at 300 nm

| mg humic acid/1 | Absorbance |
|-----------------|------------|
| 0.45 | 0.012 |
| 0.91 | 0.028 |
| 1.82 | 0.055 |
| 3.64 | 0.115 |
| 7.28 | 0.225 |
| 14.57 | 0.453 |

Practically identical values were obtained with solutions of oxidized lignin from wood hydrolysis residues.

Bohnsack[31] has described the determination of humic acids in 0.1 N NaOH solution at 420 nm.

*Gravimetric determination with lead nitrate according to Obenaus and Mücke*[240a]. The water sample is evaporated as described below, taken up in 5 ml water and filtered; the insolubles are washed. The solution is precipitated with 3 ml 0.01 M $Pb(NO_3)_2$ solution. After standing for 2 hr, the solution is decanted, the precipitate is aspirated through the weighed fritted glass vacuum filter described below, washed twice with 2 ml portions of water, dried at 105° and weighed.

For determining the lead content, the lead humate is boiled with 5 ml saturated potassium persulfate solution for 10 min, the pH of the bleached solution is adjusted to 5-6 with 0.2 N acetic acid or Urotropin, a pinch of a mixture of xylenol orange-potassium nitrate (1:100) is added and the violet solution is titrated with 0.001 M EDTA solution (Titriplex III) up to the end point with a change to lemon yellow.

1 ml 0.001 M EDTA solution = 0.21 mg Pb.

*Gravimetric determination with iron(III) chloride.* According to Obenaus[240] the water sample is evaporated to dryness on a water bath, the evaporation residue is taken up in 5 ml water, filtered, washed twice with 2 ml portions of water, treated with an addition of 1 ml sodium acetate buffer (pH = 4.8) and with 0.01 M $FeCl_3$ solution up to turbidity. After formation of the precipitate, the completeness of precipitation is tested with another drop of $FeCl_3$. The precipitate is collected in a weighed glass vacuum microfilter 12 D G 4 of Schott Co., washed with 5 ml water, dried at 105° and weighed.

To determine the iron content, the precipitate is brought into solution by alternating treatment with 1% ammonia and 10%

sulfuric acid, is evaporated to dryness with 5 ml of saturated potassium persulfate solution, the residue is fused and ashed. It is taken up with 3 x 5 ml standard acetate buffer (pH = 4.6), transferred into a 25 ml graduated flask, treated with an addition of 1 ml sodium sulfite solution (5%) and 2 ml 2,2′-dipyridyl solution (0.5%) and brought to the mark with water; after 30 min the absorbance is read at 522 nm with a S 53 filter.

Sodium acetate buffer (pH = 4.8): 40 ml dilute acetic acid (12 g glacial acetic acid per 1) + 60 ml sodium acetate solution (27.2 g $CH_3COONa \cdot 3 H_2O$ per 1).

Gjessing[108] has described the separation of humic acid on Sephadex columns G 10-G 75.

# 20. LIGNIN & LIGNOSULFONIC ACIDS FROM PULP WASTE EFFLUENTS

Among all industrial production processes, pulp waste effluents by far are the greatest organic pollutants of water. For each ton of pulp produced, the same quantity of secondary wood constituents goes into solution in the form of waste. The 1968 world pulp production amounted to 61 million tons. In countries with good wood resources and a large paper and pulp industry the wastes produced in the process, calculated as soluble organic matter, exceed the total quantity of residential waste by a multiple.

Their polluting effect on water does not reside in direct toxic action but consists primarily of the oxygen demand, growth of waste-water fungus filaments (*Sphaerotilus natans*) as well as of the impairment of color, odor and taste of the water even at high dilutions. Although fish tolerate considerable concentrations of these compounds, their meat assumes an unpleasant taste.

Essentially, two processes of pulp production from wood are distinguished: the soda pulp process, also known as the sulfate or kraft process, and sulfite pulp production. Although the former represents a smaller load on rivers, since at least 95% of the organic wastes are recovered and burnt during its performance, it is not applicable everywhere because of its noxious odor which can be detected over an area of many kilometers and can hardly be avoided.

The sulfite process, which is predominantly used in Europe, yields waste effluents of which about one third consists of easily biodegradable materials (pentoses, hexoses), while about two thirds are composed of lignosulfonic acids which are difficult to attack in biological processes. Fermentation or yeast growth from the former group of compounds, which is employed in some places, therefore represents only a limited relief of the sewer. Numerous suggestions for other uses of the sulfite effluent also can cover only extremely small quantities if they are realizable at all.

The only industrial measure with some success consists of the evaporation and incineration of the effluent liquor. The common use of calcium bisulfate for wood digestion admittedly leads to buildup of scale during evaporation formed by deposition of calcium sulfate which is difficult to control industrially. It is more favorable to use magnesium bisulfite. This permits evaporation to high concentrations without difficulties and at high incineration temperatures, the organic matter decomposes with the formation of sulfur dioxide, while the ash consists primarily of magnesium oxide. The waste gases are scrubbed with a magnesium oxide suspension and yield magnesium bisulfite solution which again serves for wood digestion. Thus, not only 95-98% of the organic wastes are eliminated and used for energy production, but the inorganic chemicals (sulfur and magnesium oxide) are largely recovered and recycled. The process has been practiced successfully for years in Lenzing (Upper Austria) among other places (see, for example, Hornke[139a]).

Lignin and lignosulfonic acids can be determined by physical methods by measuring the light absorption in the visible and UV regions as well as by fluorescence measurements. Chemical methods of analysis make use of the phenolic character of the lignins. Several phenol reagents, for example, Folin-Denis reagent with phosphotungstic-molybdic acid as well as nitrous acid respond to this character. None of the mentioned analytical methods is strictly specific; consequently, the possible presence of secondary components with a similar reaction (humic acids, phenols) must be kept in mind.

*Precipitation reaction with Trypaflavine.* According to Huhn[142] quantities of more than 50 mg/l can be determined by precipitation with 0.5% Trypaflavine hydrochloride solution and turbidimetry or gravimetry.

A still more sensitive analysis (1:1,600,000) is possible by precipitation with Surfen (1,3-bis(4-amino-2-methyl-6-quinolyl)urea) (see Vogel[319]).

*Spectrophotometric determination.* The color of lignosulfonic acids is read either directly at 280 nm or in the form of chlorolignin according to Huhn[142]. For this purpose 100 ml of water sample (pH = 6.5-7.5) are treated with 2 ml chlorine water. After 15 min 0.2 ml 25% $NH_3$ solution is added and the yellow color is read at 420 nm.

*Fluorimetric determination according to Thruston*[312]. Fluorescence excitation with UV-light of 340 nm yields a fluorescence spectrum with a maximum at 395 nm. Even 1 mg/l can be detected. The fluorescence intensity depends on the origin and molecular weight of the lignosulfonate.

Baumgartner, Feldman and Gibbons[22] have reported on the fluorescence spectra of various lignins from soda pulp waste effluents.

*Determination with nitrous acid.* According to Goldschmid and Maranville[110], 3 ml 10% $NaNO_2$ solution and 3 ml 10% acetic acid are added to 150 ml of water sample. After 15 min, 6 ml 2 N $NH_3$ solution are added and the yellow-brown color is read at 430 nm against the reagent blank value. The calibration curve is constructed with sulfite waste effluent of known solids content (1-100 mg/l). The limit of detection is 1 mg/l. A similar reaction occurs with humic acids and, for example, with bark extracts.

*Determination with Folin-Denis reagent.* Kleinert and Wincor[105] describe the following procedure:

A filtered water sample of 100 ml is treated with 0.1 g Na metapohsphate to prevent turbidity from minerals, followed by the addition of 4 ml Folin-Denis reagent. After standing for 5 min, 20 ml saturated $Na_2CO_3$ solution are added; after 45 min the blue color which has formed is read in a colorimeter with a S 66 filter. The calibration curve is constructed with 1-10 mg/l of sulfite waste effluent after its evaporation to dryness. The limit of detection is 1 mg/l. Phenols produce the same reaction.

Folin-Denis reagent: 100 g sodium tungstate, 20 g phosphomolybdic acid and 50 ml 85% phosphoric acid are boiled with 750 ml water for 2 h with refluxing.

# 21. 3,4-BENZOPYRENE

Among our environmental pollutants, polycyclic aromatic hydrocarbons are considered of importance by public health experts because of their potential carcinogenic effect. Although the atmosphere or food may be expected to represent a greater risk for the ingestion of toxic quantities in the form of dust, automotive emissions and tobacco smoke (see also Leithe[187]), drinking water pollution by these compounds has recently also been considered; their presence and analysis in water have been discussed particularly by Borneff[32]. This author lists the following guideline values for the presence of carcinogenic polycyclic hydrocarbons of various origins in water:

| | |
|---|---|
| Ground water | 1— 10 $\mu g/m^3$ |
| Clean or sparingly polluted river or lake water | 10— 50 $\mu g/m^3$ |
| Moderately polluted surface water | 50— 100 $\mu g/m^3$ |
| More highly polluted surface water | 100—1000 $\mu g/m^3$ |
| Waste water | Up to more than 100,000 $\mu g/m^3$ |

The concentration ranges involved, in which the members of this group of compounds are present not only in undissolved fractions but also in the filtrate and in aqueous solution, are so low that large quantities of water (for example, Borneff extracts 500 l of water with 18 l benzene) or extremely sensitive analytical methods will need to be used for their determination. Borneff and Kunte[33] treat 10 l of water with 600 ml of benzene.

A certain simplification of the method was realized by Scholz and Altmann[278] who limited themselves to the determination of 3,4-benzopyrene which is generally considered the major representative of carcinogenic polycyclic hydrocarbons. The determination is made by fluorimetry in dioxane solution after extraction with cyclohexane and enrichment by thin-layer chromatography.

Isolation: One l of water sample is vigorously stirred three times with 30 ml portions of ultrapure cyclohexane, each time for 5 min in the sampling vessel. The cyclohexane solutions are concentrated to 150 $\mu$l at 40° in a water-jet vacuum and nitrogen stream, and are applied as bands on a thin-layer plate (silica gel H suspended in 10% aqueous polyethylene glycol 1000 solution, applied and dried at 80° for 2 h). After evaporation of the cyclohexane the plate is covered with a coated plate to form a sandwich chamber and placed in the tank with a benzene-n-hexane solvent system (1 + 3) (a solution of 25-50 ng benzopyrene is run at the same time). After a development distance of 15 cm, (20-25 min), the plate is dried and chromatographed again as described above.

At the level of the benzopyrene spot detected by brief UV-irradiation the layer is scraped off and transferred into a small Soxhlet tube, pretreated with cyclohexane, and is extracted for 30 min with 8-10 ml cyclohexane. The filtered solution is concentrated as above, the residue is taken up in 3 ml dioxane and the solution is transferred into a quartz cuvette.

Reading: A Zeiss spectrophotometer PMQ II with a fluorescence attachment ZMF 4 and a Kodak Wratten filter 18a serves for fluorimetry. Excitation at 365 nm with a St.41 Hg lamp; quartz cuvettes of 1 cm optical path and 3 ml volume. Slit width 0.05-0.1 mm. The spectrum was recorded within 3 min from 400 to 480 nm by means of a wavelength drive on the monochromator of Philips compensating recorder with 5 mV full deflection and a paper feed of 20 mm/min. The band intensities at 429 nm in suitable length data in mm over the base line served for analysis. They were converted to full electronic amplification and constant slit aperture with consideration of correction terms based on blank and reference readings with a standard uranium glass (see the original paper for details).

With benzopyrene concentrations of 0.1-1000 ng in 3 ml dioxane solution, intensities of 66 mm per ng were obtained.

The operating range of the method falls between 0.1 and 1000 ng/l with consideration of suitable limits due to sources of error and the deviation is less than ± 15%. In low concentration ranges the introduction of benzopyrene from the laboratory and surrounding atmosphere must be anticipated and maintained at a minimum.

Jäger and Kassowitzkova[148] also reported on the determination of 3,4-benzopyrene in drinking water.

# APPENDICES

Specifications—Appendix A

*Standard specifications for waste water treatment methods* published by the Committee for Standard Specifications of waste water treatment processes sponsored by the Federal Ministry of Transportation, Federal Republic of Germany, 2nd Edition 1970 (excerpt).

| Waste Water | Process | mg KMnO$_4$/l | BOD$_5$ | Extractable with petroleum ether |
|---|---|---|---|---|
| Residential waste water, slaughter houses | Complete biological treatment | 100 | 25 | |
| Meat and fish meal factories | | | | |
| Animal meat utilization | Partial biological treatment | 150 | 80 | |
| Breweries, malt factories | Complete biological treatment | 80 | 25 | |
| | Partial biological treatment | 150 | 80 | |
| Compressed yeast factories | | 120 | 70 | |
| Dairies, cheese factories, canned milk factories | Complete biological treatment | 80 | 25 | |
| | Partial biological treatment | 120 | 80 | |
| Margarine, edible fat and oil factories | Chemical treatment | 120 | 80 | 20 |
| | Complete biological treatment | 80 | 25 | 10 |
| | Partial biological treatment | 120 | 80 | 20 |
| Soap factories | Complete biological treatment | 80 | 25 | |
| | Partial biological treatment | 120 | 80 | |
| Sugar factories | Partial biological treatment | 300 | 150 | |
| Rayon, cellulose products | | 100 | 30 | |
| Textile finishing | Chemical treatment | 200 | 100 | |
| Bleaching and dyeing plants | Biological treatment | 150 | 30 | |
| Tanning, fur finish- | Chemical treatment | 400 | 200 | 10 |
| Canning plants | Biological treatment | 150 | 80 | |
| Potato processing | Complete biological treatment | | 30 | |
| | Partial biological treatment | | 150 | |
| ing and leather plants | Complete biological treatment | | 30 | 5 |
| | Partial biological treatment | | 150 | 5 |

| Starch factories | Complete biological treatment | | 30 | |
|---|---|---|---|---|
| | Partial biological treatment | | 150 | |
| Paper and paper | Chemical treatment | 150-400 | 50-300 | |
| board factories, | Biological treatment | 100-200 | 25-40 | |
| depending on raw | | | | |
| materials (pulp, | | | | |
| wood chips, re- | | | | |
| cycled paper, rags) | | | | |
| Petroleum industry | Biological treatment | Phenols | 30 | 5 |
| | | 0.5 mg/l | | |
| Galvanizing and | | Cyanides | | |
| tempering plants | | (decom- | | |
| | | posed by | | |
| | | $Cl_2$) | | |

*Table for determination of the oxygen saturation* as a function of temperature at a total pressure of 760 torr of the water-vapor saturated atmosphere according to Truesdale, Downing and Lowden[314] (DEV G 2)

| $t_o$C | 0.0 mg $O_2$/l | 0.2 mg $O_2$/l | 0.4 mg $O_2$/l | 0.6 mg $O_2$/l | 0.8 mg $O_2$/l |
|---|---|---|---|---|---|
| 0 | 14.16 | 14.08 | 14.00 | 13.93 | 13.85 |
| 1 | 13.77 | 13.70 | 13.63 | 13.55 | 13.48 |
| 2 | 13.40 | 13.33 | 13.26 | 13.19 | 13.12 |
| 3 | 13.05 | 12.98 | 12.91 | 12.84 | 12.77 |
| 4 | 12.70 | 12.64 | 12.57 | 12.51 | 12.44 |
| 5 | 12.37 | 12.31 | 12.25 | 12.18 | 12.12 |
| 6 | 12.06 | 12.00 | 11.94 | 11.88 | 11.82 |
| 7 | 11.76 | 11.70 | 11.64 | 11.58 | 11.52 |
| 8 | 11.47 | 11.41 | 11.36 | 11.30 | 11.25 |
| 9 | 11.19 | 11.14 | 11.08 | 11.03 | 10.98 |
| 10 | 10.92 | 10.87 | 10.82 | 10.77 | 10.72 |
| 11 | 10.67 | 10.62 | 10.57 | 10.53 | 10.48 |
| 12 | 10.43 | 10.38 | 10.34 | 10.29 | 10.24 |
| 13 | 10.20 | 10.15 | 10.11 | 10.06 | 10.02 |
| 14 | 9.98 | 9.93 | 9.89 | 9.85 | 9.81 |
| 15 | 9.76 | 9.72 | 9.68 | 9.64 | 9.60 |
| 16 | 9.56 | 9.52 | 9.48 | 9.45 | 9.41 |
| 17 | 9.37 | 9.33 | 9.30 | 9.26 | 9.22 |
| 18 | 9.18 | 9.15 | 9.12 | 9.08 | 9.04 |
| 19 | 9.01 | 8.98 | 8.94 | 8.91 | 8.88 |
| 20 | 8.84 | 8.81 | 8.78 | 8.75 | 8.71 |
| 21 | 8.68 | 8.65 | 8.62 | 8.59 | 8.56 |
| 22 | 8.53 | 8.50 | 8.47 | 8.44 | 8.41 |
| 23 | 8.38 | 8.36 | 8.33 | 8.30 | 8.27 |
| 24 | 8.25 | 8.22 | 8.19 | 8.17 | 8.14 |
| 25 | 8.11 | 8.09 | 8.06 | 8.04 | 8.01 |
| 26 | 7.99 | 7.96 | 7.94 | 7.91 | 7.89 |
| 27 | 7.86 | 7.84 | 7.82 | 7.79 | 7.77 |
| 28 | 7.75 | 7.72 | 7.70 | 7.68 | 7.66 |
| 29 | 7.64 | 7.61 | 7.59 | 7.57 | 7.55 |
| 30 | 7.53 | 7.51 | 7.48 | 7.46 | 7.44 |

Dependence on prevailing atmospheric pressure:

$O_2$-saturation at $a$ torr $= O_2$-saturation at 760 torr $x \dfrac{a}{760}$

Oxygen saturation over pure oxygen:

1 l of water dissolves 44 mg $O_2$ at 20°,
1 l of water dissolves 48.5 mg $O_2$ at 15°.

## *Legislation and Government Regulations—Appendix B*

Since approximately the middle of the last century, the continuously increasing density of urban populations plus the increased water pollution due to trade and new industries have created the need to protect public waters by governmental measures. The objective is to prevent hazards to public health by water pollution as well as to avoid a deterioration of quality and losses in usability of the waters (see also Dornheim[68].

In the Federal Republic of Germany, the Water Conservation Law of 1957, effective March 1, 1960, regulates the use of public waters and provides the measures necessary for their protection. In accordance with the federal policy, this is a general law which must be made specific by suitable legislation of the countries but nevertheless contains general guidelines in order to preserve legal uniformity. The use of a public body of water requires a permit from the competent local water department. Such a permit may be connected with obligations and conditions, and it is even possible to limit or prohibit existing use privileges. Conservation regulations may be issued as ordinances and administrative orders and water-conservation guidelines may be formulated. Moreover, the laws regulate water control as well as detail questions, such as the storage of water pollutants.

Special laws and regulations connected with water conservation are the Detergent Act (see in the following), regulations controlling inland shipping, highway traffic, potable water treatment, regulations of the Food Act, and many others.

Similar laws in other German-language countries are the Law for the Protection, Utilization and Conservation of Water of 4-17-1963 of the German Democratic Republic, the Regulations for the Water Conservation Act of 1959 in Austria, and the Federal Law on Protecting Water from Pollution of 3-16-1955 of Switzerland.

## *Books and Periodicals—Appendix C*

Among books dealing with water analysis as a whole, the following may be cited:

### *German-language books*

German Standard Methods of Water Analysis (DEV). 1st Edition, Weinheim 1954; 3rd Edition, Weinheim 1968 and 1971.

K. Holl: Water, Analysis, Evaluation, Treatment. 5th Edition. Walter de Gruyter, Berlin 1970.

Handbook of Food Chemistry. Vol. VIII/1 (Water and Air). Springer-Verlag, Berlin 1969.

Investigation of Water (Plant Brochure), E. Merck, Darmstadt.

R. K. Freier: Water Analysis. Walter de Gruyter, Berlin 1964.

Klut-Olszewski: Investigation of Water. 9th Edition. Springer-Verlag, Berlin 1945.

Ohlmuller-Spitta: Investigation of Water. 4th Edition. Springer-Verlag, 1921.

J. Tillmans: Chemical Investigation of Water and Waste Water. 2nd Edition. Verlag W. Knapp. Halle 1932.

*English Language Books*

Standard Methods for the Examination of Water and Wastewater. 13th Edition. American Public Health Association, New York 1970.

ASTM Standards, Vol. 23 (Water; Atmospheric Analysis). Philadelphia 1968.

Kolthoff/Elving/Stross (Ed.): Treatise on Analytical Chemistry. Part III/2. Wiley Interscience, New York 1971.

British Ministry of Housing and Local Government (Ed.): Methods of Chemical Analysis as Applied to Sewage and Sewage Effluents. Her Majesty Stationery Office, London 1956.

Evans: Ozone in Water and Wastewater Treatment. Ann Arbor Science Publishers, Ann Arbor 1972.

Mancy: Instrumental Analysis for Water Pollution Control. Ann Arbor Science Publishers, Ann Arbor 1971, 1972, 1973.

Dorfner: Ion Exchangers Properties and Appications. Ann Arbor Science Publishers, Ann Arbor 1972.

*Periodicals*

Periodicals which are devoted especially to water analysis do not exist. Papers on the subject can be found in periodicals covering chemical analysis in general, among which we may list *Zeitschrift für analytische Chemie,* Springer-Verlag, Berlin/Heidelberg (edited by Fresenius) in Germany, *Analytical Chemistry,* American Chemical Society, Washington (USA), *Analyst* (England), Heffer, Cambridge, as well as *Analytica Chimica Acta,* Elsevier Publ. Comp., Amsterdam, and *Mikrochimica Acta,* Springer-Verlag, Vienna/New York.

Abstracts of publications on water analysis can be found in Zeitschrift fur analytische Chemie, Chemischen Zentralblatt, Verlag Chemie, Weinheim, *Chemical Abstracts,* Amer. Chem. Soc., Washington, and in *Analytical Abstracts,* Fisk, London. The reader may also refer to the "Literature Reports on Water, Waste Water, Air and Solid Waste," edited by Meinck (Berlin). *Analytical Chemistry* publishes an issue of abstracts every two years which also contains a chapter with references and citations concerning advances in water analysis.

Moreover, water analysis is covered in various water and waste water periodicals. In such periodicals this area is covered together with that of gases. This is done either under the main title of environmental protection or with consideration of the fact that the activity of many chemists in public service and industry deals with water and gas generation at the same time. Some frequently cited periodicals with these orientations are the following:

*Germany*

Archiv für Hygiene und Bakteriologie, Verlag Urban & Schwarzenberg Munich/Vienna.

Gas- und Wasserfach (Wasser/Abwasser), Verlag Oldenbourg, Munich.

Gesundheits-Ingenieur, Verlag Oldenbourg, Munich.
Korrespondenz Abwasser, Verlag Ges. zur Förderung der Abwassertechnik, Bonn.
Vom Wasser (Yearbook), Verlag Chemie, Weinheim.
Städtehygiene, Neuer Hygiene-Verlag, Uelzen.
Wasser/Luft/Betrieb, Krausskopf-Verlag, Mainz.
Wasserwirtschaft/Wassertechnik, VEB Verlag für Bauwesen, Berlin.
Wasser- und Abwasserforschung, Wasser- und Abwasserforschungs-Verlagsges., Munich.
Fortschritte der Wasserchemie und Grenzgebiete, Akademieverlag, Berlin.

*Austria*

Gas/Wasser/Wärme, Österr. Verein für das Gas- und Wasserfach, Vienna.
Österreichische Wasserwirtschaft, Springer-Verlag Vienna/New York.
Österr. Abwasser-Rundschau, Verlag Österr. Abwasserrundschau, Vienna.
Umweltschutz-Städtereinigung, Bohmann Industrie- und Fachverlag, Vienna.
Wasser und Abwasser (Yearbook), Verlag Winkler, Vienna.

*Switzerland*

Schweizerische Zeitschrift für Hydrologie, Birkhäuser-Verlag, Basel.

*USSR*

Gigiena y Sanita, Verlag Medgis, Moscow.

*England and USA*

Air and Water Pollution, Pergamon Press, Oxford.
Environmental Science and Technology, Amer. Chem. Society, Washington.
Journal of the American Water Works Association, Washington.
Journal of the Water Pollution Control Federation, Washington.
Public Health, Society of Medical Officers of Health, London.
Water and Sewage Works, Scranton Publishing Comp., Chicago.
Water Research, Pergamon Press, London.

# REFERENCES

1   Abbott, D. C.: Analyst 87, 286 (1962).
2   Abbott, D. C.: Analyst 88, 240 (1963).
3   Abbott, D. C., J. A. Bunting and J. Thomson: Analyst 90, 356 (1965).
4   Abbott, D. C., H. Egan, E. Hammond and E. W. Thomson: Analyst 89, 480 (1964).
5   Abrahamczik, E., G. Groh, W. Huber and F. Kraus: Vom Wasser 37, 82 (1970).
5a  Adelmann, M. H.: Water/Wastes Engin. 5, 52 (1968).
6   Ahling, B. and J. Jensen: Anal. Chemistry 42, 1483 (1970).
7   Albersmeyer, W.: Stadtehyg. 5, 85 (1958); Gas-Wasserfach 99, 269 (1958).
8   Aly, O. M.: Water Research 2, 587 (1968).
9   Ammon, F. v.: Vom Wasser 23, 162 (1961).
10  An der Lan, H.: Wasser and Abwasser 1965, 85.
11  Argauer, R. J.: Anal. Chemistry 40, 122 (1968).
12  Armstrong, F. A. J. and G. T. Boalch: Nature 192, 858 (1961).
13  Askew, J., J. H. Ruzicka and B. B. Wheals: Analyst 94, 275 (1969).
13a Asmus, E. and H. Garschagen: Z. anal. Ch. 138, 414 (1953).
14  Axt, G.: Vom Wasser 36, 328 (1969).
15  Bahensky, V. and Z. Zika: Z. Galvanotechnik 53, 122 (1962).
15a Baker, R. A.: J. Amer. Water Works Assoc. 58, 751 (1966).
16  Baker, R. A.: Air Water Poll. 10, 591 (1966); Water-Research 1, 61, 97 (1967); 3, 717 (1969); 4, 559 (1970); J. Water Poll Control Fed. 37, 1164 (1965).
17  Baker, R. A.: J. Amer. Water Works Assoc. 58, 751 (1966); Air Water Poll. 10, 591 (1966); Envir. Sci. Technol. 1, 997 (1967).
18  Baker, R. A. and B. A. Malo: Water Sewage Works 114, R 223 (1967).
18a Banyon, L. R., R. Kaschnitz and G. W. A. Rijnders: Stichting Concawe, Dec. 1968.
19  Bark, L. S. and H. C. Higson: Talanta 11, 621 (1964).
19a Bauer, K. and H. Driescher: Fortschr. Wasserchem. 10, 31 (1968).
20  Bauer, L. and W. Schmitz: Vom Wasser 36, 383 (1969).
20a Bauer, L.: Vom Wasser 38, 49 (1971).
21  Baughman, G. L., B. T. Butler and W. M. Sanders: Water Sewage Works 116, 359 (1969).
22  Baumgartner, D. J., M. H. Feldman and C. L. Gibbon: Water Research 5, 533 (1971).
23  Baumler, J. and S. Rippstein: Helv. Chim. Acta 44, 1162 (1961).
24  Beran, F. and J. A. Guth: Pflanzenschutzber. 33, Heft 5-6; Wasser and Abwasser 1965, 57.
25  Berck, B.: Anal. Chemistry 25, 1253 (1953).
25a Berthold, J.: Z. anal. Ch. 240, 320 (1968).
26  Beuthe, C. G.: Ges.-Ing. 83, 70 (1962); Gas- u. Wasserfach 111, 689 (1970).
27  Bevenue, A., T. W. Kelly and J. W. Hylin: J. Chromatogr. 54, 71 (1871).

27a Bevenue, A. and J. N. Ogada: J. Chromatogr. 50, 142 (1970).

28   Blaedel, W. J., D. B. Easty, L. Anderson and T. R. Farrell: Anal. Chemistry 43, 890 (1971).

29   Blinn, R. C.: Residue Reviews 5, 130 (1964).

30   Blinn, R. C. and F. A. Gunther: Residue Reviews 2, 99 (1963).

31   Bohnsack, G.: Mitt. Ver. Großkesselbetr. 73, 276 (1961).

32   Borneff, J.: Gas-Wasserfach 108, 1072 (1967); Arch. f. Hygiene 148, 585 (1964).

33   Borneff, J. and H. Kunte: Arch. f. Hygiene 153, 220 (1969).

34   Boye, E.: Z. anal. Ch. 207, 260 (1964).

35   Braun, B.: Gas-Wasserfach 111, 663 (1970).

36   Brauss, F. W. and H. Barth: Arch. f. Hygiene 153, 467 (1969).

37   Breyer, B., and H. H. Bauer: Alternating Current Polarography and Tensammetry, New York 1963.

38   Bridie, A. L. A. M.: Water Research 3, 157 (1969).

39   Brink, D. W.: Analyst 87, 193 (1962).

40   Browns, E. G. and T. J. Hayes: Analyst 80, 755 (1955).

40a Bucksteeg, W. and F. Dietz: Vom Wasser 24, 282 (1957).

41   Bucksteeg, W. and F. Dietz: Wasser/Luft/Betr. 13, 417, 475 (1969).

42   Bucksteeg, W. and H. Thiele: Gas/Wasserfach 98, 26 (1957).

43   Bunyan, P. J.: Analyst 89, 615 (1964).

44   Burchardt, C. H.: Gas/Wasserfach 107, 42 (1966); 109, 1189 (1968).

44a Burchfield, H. P. and R. J. Wheeler: J. Assoc. Off. Anal. Ch. 49, 651 (1966).

45   Burger, K.: Z. anal. Ch. 196, 251 (1963).

45a Burmeister, H. E.: Munchner Beitr. 9, 213 (1967).

46   Burns, E. R. and C. Marshall: J. Water Poll. Contr. Fed. 37, 1716 (1965).

47   Casapieri, P., R. Scott and E. A. Simpson: Anal. Chim. Acta 49, 188 (1970).

48   Challacombe, J. A. and J. A. McNulty: Residue Reviews 5, 57 (1964).

49   Clark, J. W.: Water Sewage Works 107, 140 (1960).

50   Clark, S. J.: Residue Reviews 5, 32 (1964).

51   Claeys, R. R. and H. Freund: Envir. Sci. Technol. 2, 458 (1968).

52   Clifford, D. A.: Proceed. 23rd Ind. Waste Conf. Purdue Univ. 1968, 772.

53   Cohen, J. C., J. Norcup and J. H. A. Ruzicka: J. Chromatogr. 49, 215 (1970).

54   Coulson, D. M. and L. A. Cavanagh: Anal. Chemistry 32, 1245 (1960).

55   Coulson, D. M., L. A. Cavanagh, J. E. De Vries and B. J. Walter: J. Agr. Food Chem. 8, 399 (1960).

56   Courtot-Coupez, J. and A. le Bihan: Chem. Abstr. 71, 15847w (1969).

57   Crabb, N. T. and H. E. Persinger: J. Amer. Oil Chem. Soc. 41, 752 (1964).

58   Cripps, J. M. and D Jenkins: J. Water Poll. Contr. Fed. 36, 1240 (1964).

59   Cropper, F. R., D. M. Heinekey and A. Westwall: Analyst 92, 436, 443 (1967).

60   Cropper, F. R. and D. M. Heinekey: Analyst 94, 484 (1969).

61   Crosby, N. T. and E. O. Laws: Analyst 89, 319 (1964).

61a Daniel, R. L. and R. B. Le Blanc: Anal. Chemistry 31, 1221 (1959).

62   Dacre, J. C.: Anal. Chemistry 43, 589 (1971).

62a Dietrich, K. R.: Waste water technology. Heidelberg 1968.

63  Dietz, F. and P. Koppe: Vortr.-Ber. Haus d. Techn. (Essen) 1971 (im Druck).
64  Dimick, K. P. and H. Hartmann: Residue Reviews 5, 150 (1964).
65  Dobbs, R. A. and R. T. Williams: Anal. Chemistry 35, 1064 (1963).
66  Dobbs, R. A., R. H. Wise and R. B. Dean: Anal. Chemistry 30, 1255 (1967).
67  Dohrmann (Fa.): Application Series A I 10.
68  Dornheim, C.: Handbook of Food Chemistry VIII/2, pp. 1228-41.
69  Drescher, N.: Vom Wasser 34, 224 (1967).
70  Dyatlovitskaya, F. G. and E. F. Gladenko: Gig. y Sanit. 33, 53 (1968); Z. anal. Ch. 250, 62 (1970).
71  Ecker, E.: Ch. Ztg. 95, 511 (1971).
72  Egan, H., E. W. Hammond and J. Thomson: Analyst 89, 175 (1964).
73  Egli-Schar, H.: Z. Wasser- u. Abw.-Forschg. 1, 83 (1968).
74  Eichelberger, J. W., R. C. Dressman and J. E. Longbottom: Envir. Sci. Technol. 4, 576 (1970).
75  El-Dib, M. A.: J. Assoc. Off. Anal. Chem. 53, 756 (1970).
76  Emmet, R. T.: Anal. Chemistry 41, 1648 (1969).
77  Epstein, J.: Anal. Chemistry 19, 272 (1947).
77a Erichsen, L. v. and N. Rudolphi: Erdol u. Kohle 8, 16 (1955).
78  Evans, W. H.: Analyst 87, 569 (1962).
79  Fairing, J. D. and F. R. Short: Anal. Chemistry 28, 1827 (1956).
80  Farrow, R. N. P. and A. G. Hill: Analyst 90, 241 (1965).
81  Fastabend, W.: Dechema-Monogr. 52, 85 (1964).
82  Fastabend, W. and M. Handloser: Vom Wasser 33, 142 (1966).
83  de Faubert Maunder, M. J., H. Egan and J. Roburn: Analyst 89, 157, 168 (1964).
84  Faust, S. D. and E. W. Mikulewicz: Water Research 1, 509 (1967).
85  Fischer, R. and W. Klingelholler: Arch. Toxikol. 19, 119 (1961).
86  Fischer, W. K.: Vom Wasser 32, 168 (1965).
87  Fischer, W. K.: Munchner Beitr. Abwasserforschg. 9, 24 (1967).
88  Fleet, B. and H. v. Storp: Anal. Chemistry 43, 1575 (1971).
89  Fleps, W.: Ges. Ing. 84, 209 (1963).
90  Formaro, L. and S. Trasatti: Anal. Chemistry 40, 1060 (1968).
91  Foulds, J. M. and J. V. Lunsford: Water Sewage Works 115, 112 (1968).
92  Frank, P. A., and R. J. Demint: Envir. Sci. Technol. 3, 69 (1969).
93  Freimuth, U., et al.: Fortschr. Wasserchem. 11, 189 (1969).
94  Friedrichs, J.: Ch. Ztg. 55, 519 (1931).
95  Froboese: Arb. Reichsgesundheitsamt 52, 211 (1920).
96  Fuchs, W. and E. Kohler: Mitt. Ver Großkesselbetr. 47, 107 (1957).
97  Fuchs, J.: Chem. Ztg. 84, 423, 805 (1960).
98  Fuhrmann, D. L., G. W. Latimer and J. Bishop: Talanta 13, 103 (1966).
99  Fuhs, G. W.: Wasser- u. Abw.-Forschg. 1, 161 (1968).
100 Gad, G. and H. Schlichting: Ges. Ing. 76, 373 (1955).
101 Gaston, L. K.: Residue Reviews 5, 21 (1964).
102 Gertner, A. and H. Ivecovich: Z. anal. Ch. 142, 36 (1954).
103 Getz, M. E.: Residue Reviews 2, 9 (1963).
104 Giang, P. A. and S. A. Hall: Anal. Chemistry 23, 1830 (1951).
105 Giebler, G., P. Koppe and H. T. Kempf: Gas/Wasserfach 105, 1039, 1093 (1964).

106    Giuffrida, L. and F. N. Ives: J. Assoc. Off. Anal. Chemists 52, 541 (1969).
107    Gjavotchanoff, St., H. Lussem and E. Schlimme: Gas/Wasserf. 112, 448 (1971).
108    Gjessing, E. T.: Chem. Abstr. 70, 40575 (1969).
108a   Goebgen, H. G.: Haus d. Techn. (Essen) Vortrags-Veroff. 231, 55 (1970).
109    Goebgen, M. G. and J. Brockmann: Wasser/Luft/Betr. 13, 204 (1969).
110    Goldschmid, O. and L. Fr. Maranville: Anal. Chemistry 31, 370 (1959).
111    Goodwin, E. S., R. Goulden and J. G. Reynolds: Analyst 86, 697 (1961).
112    Gorbach, G. and F. Ehrenberger: Z. anal. Ch. 181, 106 (1961).
113    Gorbach, G., O. G. Koch and G. Dedic: Mikrochim. Acta 1955, 882.
114    Gordon, G. E.: Anal. Chemistry 32, 1325 (1960).
115    Greff, R. A., S. A. Setzkorn and W. D. Leslie: J. Amer. Oil Chem. Soc. 42, 180 (1965).
116    Greff, R A., E. A. Setzkorn and W. D. Leslie: Anal. Abstr. 13, 3650 (1966).
117    Grutz, P. W. E.: Water Research 1, 319 (1967).
118    Gunther, K.: Fortschr. Wasserchem. 10, 107 (1968).
119    Gunther, F. A. and R. C. Blinn: Analysis of Insecticides, Vol. VI, New York 1955, p. 231.
120    Habermann, J. P.: Anal. Chemistry 43, 63 (1971).
121    Hall, C. E. van, J. Safranko and V. A. Stenger: Anal. Chemistry 35, 315 (1963).
122    Hall, C. E. van, D. Barth and V. A. Stenger: Anal. Chemistry 37, 769 (1965).
123    Hall, C. E. van and V. A. Stenger: Anal. Chemistry 39, 503 (1967).
124    Hamilton, D. J. and B. W. Simpson: J. Chromatogr. 39, 186 (1969).
125    Hancock, W. and E. A. Laws: Analyst 80, 665 (1955).
126    Harwood, J. E. and D. J. Huyser: Water Research 2, 631 (1968).
126a   Headington, C. E. and 10 Mitarbeiter: Anal. Chemistry 25, 1681 (1953).
127    Heertjes, P. M. and A. P. Meijers: Gas/Wasserfach 111, 61 (1970).
128    Heier, H.: Fortschr. Wasserchem. 12, 20 (1970).
129    Heinerth, E.: Tenside 3, 109 (1966).
130    Heinz, H. J. and W. K. Fischer: Fette/Seifen/Anstrichm. 66, 685 (1964).
131    Hellmann, H.: Z. anal. Ch. 244, 44 (1969).
132    Hellmann, H.: Dtsch. Gewasserkundl. Mitt. 13, 19 (1969).
133    Herzel, F.: Ges. Ing. 88, 379 (1967).
133a   Herzel, F.: Vom Wasser 38, 71 (1971).
134    Hettche, H. O.: Ges Ing. 76, 309 (1955); Gas/Wasserfach 96, 660 (1955).
135    Hey, A. E., A. Green and N. Harkness: Water Research 3, 873 (1969).
136    Hey, A. E. and S. H. Jenkins: Water Research 3, 887 (1969).
137    Hiser, L. L. and A. W. Busch: J. Water Poll. Contr. Fed. 36, 505 (1964).
138    Hissel, J. and M. Cadot-Dethier: Trib. CEBEDEAU 18, 272 (1965).
139    Holluta, J. and K. Hochmuller: Vom Wasser 26, 146 (1959).
139a   Hornke, R.: Wasser and Abwasser 1965, 206.
140    Huddleston, R. L. and R. D. Allred: J. Amer. Oil Chem. Soc. 42, 983 (1965).
141    Huditz, K. and H. Flaschka: Z. anal. Ch. 136, 185 (1952).
142    Huhn, W.: Fortschr. Wasserchem. 1, 95 (1964).
143    Husmann, W., F. Malz and H. Jendreyko: Research Reports of Nordrhein-Westfalen No. 1153 (1963).
144    Huygsten, J. J. van: Water Research 4, 645 (1970).

144a Imhoff, K.: Pocketbook of Municipal Drainage Technology. 22nd Ed., **Munich, 1968.**
145 Irmann, F.: Ch. Ing. Techn. 37, 789 (1965).
146 Janicke, W.: Gas/Wasserfach 109, 246 (1968).
147 Janicke, W.: Fette/Seifen/Anstrichm. 71, 843 (1969).
148 Jager, J. and B. Kassowitzkowa: Chem. Listy 62, 216 (1968): Z. anal. Ch. 247, 363 (1969).
149 Jeltes, R.: Water Research 3, 931 (1969).
150 Jeltes, R. and R. Veldink: J. Chromatogr. 27, 242 (1967).
151 Jenkins, S. H., A. E. Hey and J. S. Cooper: Int. J. Air Water Poll. 10, 495 (1966).
152 Jeris, J. S. A.: Water/Wastes Engineering 4, 89 (1967).
153 Johns, Th. and Ch. H. Braithwaite: Residue Reviews 5, 45 (1964).
154 Kahn, L. and C. H. Wayman: Anal. Chemistry 36, 1340 (1964).
155 Kaiser, R.: Chromatographia 1, 199 (1968); J. Chrom. Sci.
156 Kaiser, R.: Vortragsveroff. 1971 (Haus d. Techn. Essen).
157 Kammerer, P. A. and G. F. Lee: Envir. Sci. Technol. 3, 276 (1969).
158 Katz, J., S. Abraham and N. Baker: Anal. Chemistry 26, 1503 (1954).
159 Kawahara, F. K.: Anal. Chemistry 40, 1009 (1968).
160 Kawahara, F. K., J. J. Lichtenberg and J. W. Eichelberger: J. Water Poll. Contr. Fed. 39, 446 (1967).
161 Kawahara, F. K., R. L. Moore and R. W. Gorman: J. Gaschrom. 6, 24 (1968).
162 Knappe, E. and I. Rohdewald: Z. anal. Ch. 200, 9 (1964).
163 Kempf, Th.: Ges. Ing. 81, 169 (1960).
164 Kieselbach, R.: Anal. Chemistry 26, 1312 (1954).
165 Kleinert, Th. and W. Wincor: Papier 6, 513 (1952).
166 Klotter, H. E. and E. Hantge: Wasserwirtsch. 56, 21 (1966).
167 Kneip, T. J., T. H. Beasley, P. King and W. K. Dean: Anal. Chemistry 39, 1510 (1967).
168 Kolaczkowski, St., Z. Mejboum and S. Spandowska: Fortschr. Wasserchem. 7, 195 (1967).
168a Kolle, W.: Vom Wasser 36, 34 (1969).
169 Kolle, W. and H. Sontheimer: Gas/Wasser/Warme 22, 226 (1968).
170 Konig, H.: Z. anal. Ch. 251, 167 (1970).
171 Konig, H.: Z. anal. Ch. 251, 359 (1970).
172 Konrad, J. G., H. B. Pionke and G. Chesters: Analyst 94, 490 (1969).
173 Koppe, P. and K. Muhle: Vom Wasser 35, 42 (1968).
174 Koppe, P. and C. Pittag: Gas/Wasserfach 107, 1311, 1424 (1966).
175 Koppe, P. and I. Rautenberg: Wasser/Luft/Betr. 14, 419 (1970).
176 Koppe, P. and I. Rautenberg: Gas/Wasserfach 111, 80 (1970).
177 Kratochwil, V.: Coll. Cech. Chem. Comm. 25, 299 (1960).
178 Krieger, H.: Gas/Wasserfach 104, 695 (1963).
179 Kupchik, G. J. and G. P. Edwards: J. Water Poll. Contr. Fed. 34, 376 (1962).
179a Kussmaul, H. and A. Muhle: Vortragsveroff. Haus d. Tech. Essen 1971.
180 Ladenburg, P.: Vom Wasser 29, 119 (1962).
181 Lawerenz, A.: Fortschr. Wasserchem. 10, 21 (1968); Z. ges. Hyg. 12, 391 (1966).
182 Le Bihan, A. and J. Courtat-Coupez: Bull. Soc. Chim. France 1970, 406.
183 Lee, E. G. H. and C. C. Walden: Water Research 4, 641 (1970).

184  Leibnitz, E., U. Behrens, H. Koll and H. Richter: Chem. Techn. 14, 33 (1962).
185  Leithe, W.: Z. Unters. Lebensm. 67, 441 (1934).
186  Leithe, W.: Z. anal. Ch. 193, 16 (1963).
187  Leithe, W.: Analysis of the Atmosphere and Its Pollutants. Stuttgart 1968.
188  Leithe, W.: Gas/Wasserfach 110, 1233 (1969).
189  Leithe, W.: Wasser/Luft/Betr. 14, 220 (1970).
190  Leithe, W.: Chemie/Labor/Betr. 21, 433 (1970).
191  Leithe, W.: Vom Wasser 37, 106 (1970).
192  Leithe, W.: Osterr. Abwasser-Rundschau 15, 25 (1970).
193  Leithe, W.: Wasser/Luft/Betr. 14, 55 (1970).
194  Leithe: W.: Z. anal. Ch. 251, 185 (1970).
195  Leithe, W.: Osterr. Abwasser-Rundsch. 16, 4 (1971).
196  Leithe, W.: Ch. Ztg. 95, 452 (1971).
197  Leithe, W.: Ch. Ztg. 95, 463 (1971).
198  Leithe, W.: Correspondence, Abwasser 18, 126 (1971).
199  Leithe, W.: Vom Wasser 38, 119 (1971).
199a Leithe, W. and G. Petschl: Z. anal. Ch. 230, 344 (1967).
200  Leschber, R.: Galvanotechnik 58, 462 (1967).
200a Leschber, R. and W. Niemitz: Ges. Ing. 90, 229 (1969).
201  Leschber, R. and H. Schlichting: Z. anal. Ch. 245, 300 (1969).
202  Levine, W. S., G. S. Mapes and M. J. Roddy: Anal. Chemistry 25, 1840 (1953).
203  Liebmann, H. and K. Offhaus: Abwassertechnik 17, IV-VI (1966).
204  Longwell, J. and W. D. Maniece: Analyst 80, 167 (1955).
205  Lovell, V. M. and F. Sebba: Anal. Chemistry 38, 1926 (1966).
206  Ludemann, D.: Handbook of Food Chemistry, VIII/2, pp. 1193-1199.
207  Ludemann, D. and H. Neumann: Z. angew. Zool. 47, 11, 303 (1960).
208  Ludzak, F. J., W. A. Moore and C. C. Ruchhoft: Anal. Chemistry 26, 1784 (1954).
208a Ludzak, F. J. and C. E. Whitfield: Anal. Chemistry 28, 157 (1956).
209  Lussem, H.: Haus d. Techn. Vortragsveroff. 231, 88 (1970).
210  Lurje, J. J.: Fortschr. Wasserchem. 7, 81 (1967).
211  Lysyi, J. and K. H. Nelson: J. Gas-Chrom. 6, 106 (1968); Anal. Chemistry 40, 1365 (1968).
212  Lysyi, J., K. H. Nelson and P. P. Newton: J. Water Poll. Contr. Fed. 40, R 181 (1968); Water Research 2, 233 (1968).
213  Lysyi, J., K. H. Nelson and St. R. Webb: Water Research 4, 157 (1970).
214  Maehler, C. Z., J. M. Cripps and A. E. Greenberg: J. Water Poll. Contr. Fed. 39, R 92 (1967).
215  Maier, D.: Vom Wasser 36, 246 (1969).
215a Malissa, H.: Mikrochim. Acta 1957, 553; 1960, 127.
216  Malz, F. and J. Gorlas: Vom Wasser 34, 209 (1967).
217  Martin, J. M., C. R. Orr, C. B. Kincannon and J. L. Bishop: J. Water Poll. Contr. Fed. 39, 21 (1967).
218  Meinck, F., H. Stoof and H. Kohlschutter: Industrial Waste Effluents. 4th Ed., Stuttgart 1968.
219  Melpolder, F. W., C. W. Warfield and C. E. Headington: Anal. Chemistry 25, 1453 (1953).
220  Mertens, H.: Gas/Wasserfach 110, 349 (1969).

221 Milwidsky, B. M.: Analyst 94, 377 (1969).
222 Mirsch, E.: Fortschr. Wasserchem. 2, 126 (1965).
223 Mohler, E. F. and J. N. Jacob: Anal. Chemistry 29, 1369 (1957).
225 Montgomery, H. A. C.: Water Research 1, 631 (1967).
226 Montgomery, H. A. C. and A. Cockburn: Analyst 89, 679 (1964).
226a Montgomery, H. A. C., Gariner and Gregory: Analyst 94, 284 (1969).
227 Montgomery, H. A. C., J. F. Dymock and N. S. Thom: Analyst 87, 949 (1962).
227a Montgomery, H. A. C. and N. S. Thom: Analyst 87, 689 (1962).
228 Montgomery, H. A. C., N. S. Thom and A. Cockburn: J. appl. Chem. 14, 280 (1964).
229 Moore, W. A. and R. A. Colbeson: Anal. Chemistry 28, 161 (1956).
230 Moore, W. A., C. Kroner and C. C. Ruchhoft: Anal. Chemistry 21, 953 (1949).
231 Moore, W. A., F. J. Ludzak and C. C. Ruchhoft: Anal. Chemistry 23, 1297 (1951).
232 Morkowski, J.: Wasser/Luft/Betr. 2, 345 (1962).
233 Mrkva, M.: J. Water Poll. Contr. Fed. 41, 1923 (1969).
233a Muers, M. M.: Anal. Chemistry 22, 846 (1950).
234 Mullis, M. K. and E. D. Schroeder: J. Water Poll. Contr. Fed. 43, 209 (1971).
235 Murtaugh, J. J. and R. L. Bunch: J. Water Poll. Contr. Fed. 39, 404 (1967).
236 Mutter, M.: Chromatographia 2, 208 (1969).
236a Naucke, W.: Brennst. Chem. 45, 38 (1964).
237 Naucke, W. and F. Tackmann: Brennst. Chem. 45, 263 (1964).
238 Nelson, K. H. and J. Lysyi: Envir. Sci. Technol. 2, 61 (1968).
239 Niemitz, W. and K. Fuss: Gas/Wasserfach 104, 117 (1963).
239a Nietsch, B.: Vom Wasser 21, 186 (1954).
240 Obenaus, R.: Schweizer Ztschr. Hydrol. 25, 9 (1963).
240a Obenaus, R. and D. Mucke: Z. anal. Ch. 201, 428 (1964).
241 Offhaus, K. and W. Waschiczek: Wasser- u. Abw.-Forschg. 2, 56 (1969).
241a Ogden, C. P., H. L. Webster and J. Halliday: Analyst 86, 22 (1961).
242 Ogura, N. and T. Hanya: J. Water Poll. Contr. Fed. 40, 464 (1968).
242a Oehler, K. E.: In: Handbook of Food Chemistry, Vol. VIII/1, pp. 303-322.
243 Opperskalski, K. and W. Siebert: Vom Wasser 37, 92 (1970).
244 Oster, H.: Siemens-Ztschr. 42, 703 (1968); Z. anal. Ch. 247, 257 (1969).
245 Panowitz, J. F. and C. E. Renn: J. Water Poll. Contr. Fed. 38, 636 (1966).
246 Perkow, W.: Active Ingredients in Repellants and Pesticides, Berlin 1971.
246a Perry, S. G.: Chemistry in Britain 7, 366 (1971).
247 Petrowitz, H. J.: Chem. Ztg. 85, 867 (1961).
248 Pickhardt, W. P., A. N. Oemler and J. Mitchell: Anal. Chemistry 27, 1784 (1955).
249 Pionke, H. B., J. G. Konrad, G. Chesters and D. E. Armstrong: Analyst 93, 363 (1968).
250 Piorr, W.: Z. Lebensm.-Unters. u. Forschg. 132, 140 (1966).
250a Pociecha, Z.: Chem. Anal. (Warsaw) 12, 243 (1967).

251    Pollerberg, J.: Fette/Seifen/Anstrichm. 69, 179 (1967).
252    Popel, F., K. H. Hunken and H. Steinecke: Gas/Wasserfach 99, 897 (1958).
253    Puschel, R., and H. Grubitsch: Brennst. Chem. 38, 266 (1957).
254    Quentin, K. E. and E. Huschenbeth: Vom Wasser 35, 76 (1968).
255    Rather, J. B. and 10 Mitarbeiter: Anal. Chemistry 30, 36 (1958).
256    Reed, T. D.: Proceed. Soc. Anal. Chemistry 7, 32 (1970).
257    Reichert, J. K.: Gas/Wasserfach 112, 403 (1971).
258    Reploh, H. and G. Bunemann: Arch. f. Hygiene 146, 562 (1963).
259    Riley, W. P. and D. Taylor: Anal. Chim. Acta 46, 307 (1969).
260    Roberts, R. F. and B. Jackson: Analyst 96, 209 (1971).
261    Rosen, A. A. and F. M. Middleton: Anal. Chemistry 31, 1729 (1959).
261a   Rosen, A. A. and M. Rubin: J. Water Poll. Contr. Fed. 37, 1302 (1967).
262    Rubelt, C.: Z. anal. Ch. 221, 299 (1966).
263    Rubelt, C.: Vortragsveroff. Haus d. Techn. (Essen) 231, 65 (1970).
264    Rubelt, C., W. Schweisfurth and W. Zimmermann: Gas/Wasserfach 108, 893 (1967).
265    Rudling, L.: Water Research 4, 533 (1970).
266    Russell, F. R. and W. T. Wilkinson: Analyst 84, 751 (1959).
267    Sallee, E. M., et al.: Anal. Chemistry 28, 1822 (1956).
268    Salzer, F.: Microchem. J. 10, 27 (1966); Z. Anal. Ch. 205, 66 (1964).
269    Sander, P. Vortragsveroff. Haus d. Techn. (Essen) 231, 4 (1970).
270    Sanderson, W. W. and B. Ceresia: J. Water Poll. Contr. Fed. 37, 1176 (1965).
270a   Sattelmacher, P. G. and E. Furstenau: Ges. Ing. 82 (1961).
271    Schaffer, R. B., et al.: J. Water Poll. Contr. Fed. 37, 1545 (1965).
272    Schechter, M. S. and Haller: Ind. Eng. Chem. Anal. Ed. 17, 704 (1945).
273    Schechter, M. S. and J. Hornstein: Anal. Chemistry 24, 544 (1952).
274    Schmauch, L. J. and H. M. Grubb: Anal. Chemistry 26, 308 (1954).
275    Schmid, M. and K. H. Mancy: Schweiz. Ztschr. Hydrol. 32, 328 (1970).
276    Schneider, C. R. and H. Freund: Anal. Chemistry 34, 69 (1962).
277    Scholz, L.: Vom Wasser 32, 143 (1965); Z. anal. Ch. 202, 425 (1964).
278    Scholz, L. and H. J. Altmann: Z. anal. Ch. 240, 81 (1968).
279    Schuller, A. J.: Arch. Hyg. Bakteriol. 145, 201 (1961).
280    Schutzmann, R. L. and W. F. Barthel: Chem. Abstr. 70, 56345 (1969).
281    Schwarz, K.: Z. anal. Ch. 115, 161 (1939).
282    Sebban, R.: Chim. Analytique 50, 123 (1969).
282a   Seeboth, H.: Fortschr. Wasserchem. 2, 128 (1965); Chem. Techn. 15, 34 (1963).
282b   Seeboth, H. and H. Gorsch: Chem. Techn. 15, 294 (1963).
283    Segal, S. and M. L. Sutherland: Residue Reviews 5, 73 (1964).
284    Sementschenko, L. V. and V. T. Kaplin: Zavodsk. Lab. 33, 801 (1967); Anal. Abstr. 15, 3633 (1968).
285    Sheiham, J. and T. A. Pinfold: Analyst 94, 387 (1969).
286    Sierp, F.: Techn. Gemeindoblatt 30, 179 (1927).
287    Sierp, F. and Fransemeyer: Techn. Gemeindebl. 34, 233 (1931).
288    Sherrat, J. G.: Analyst 81, 518 (1956); 87, 595 (1962).
289    Sigworth, E. A.: J. Amer. Water Works Assoc. 57, 1016 (1965).
290    Simard, R. G., J. Hasegawa, W. Bandaruk and C. E. Headington: Anal. Chemistry 23, 1384 (1951).
290a   Slack, J. G.: Analyst 84, 193 (1959).

291 Smith, D. and J. Eichelberger: J. Water Poll. Contr. Fed. 37, 77 (1965).
292 Sodergren, A.: Analyst 91, 113 (1966).
293 Sontheimer, H.: Gas/Wasserfach 111, 93 (1970).
294 Sontheimer, H.: Gas/Wasserf. 111, 420 (1970).
295 Soucek, J., P. Popovska and J. Sindelar: Fortschr. Wasserchem. 11, 43 (1969).
296 Sprenger, F. J.: Vom Wasser 36, 319 (1969).
296a Stankovic, V.: Fortschr. Wasserchem. 3, 144 (1965).
297 Stenger, V. A. and C. E. van Hall: J. Water Poll. Contr. Fed. 40, 2 (1968).
298 Steveninck, J. van, and J. C. Riemsma: Anal. Chemistry 38, 1250 (1966).
299 Strackenbrock, K. H.: Ges. Ing. 79, 54 (1956).
300 Strohl, G. W.: Ges. Ing. 88, 382 (1967).
301 Strohl, G. W.: Mikrochim. Acta 1969, 130.
302 Sugar, J. W. and R. A. Conway: J. Water Poll. Contr. Fed. 40, 1622 (1968).
303 Swisher, R. D., M. M. Crutchfield and D. W. Caldwell: Envir. Sci. Technol. 1, 820 (1967).
304 Taylor, R.: Water Poll. Abstr. 42, 204 (1969).
305 Taylor, C. G. and B. Fryer: Analyst 94, 1106 (1969).
306 Taylor, R. and T. Bogacka: Chem. Anal. (Warsaw) 13, 227 (1968); Z. anal. Ch. 250, 61 (1970).
307 Thielemann, H.: Z. Chem. 9, 390 (1969).
308 Thielemann, H.: Z. Wasser- u. Abwasserforschg. 3, 122 (1970).
309 Thielemann, H.: Z. Wasser- u. Abwasserforsch. 3, 154 (1970); Z. Chem. 9, 189 (1969).
310 Thielemann, H.: Wasserwirtsch. Wassertechn. 21, 336 (1971).
311 Thompson, J. E. and J. R. Duthrie: J. Water Poll. Contr. Fed. 40, 317 (1968).
312 Thruston, A. D.: J. Water Poll. Contr. Fed. 42, 1551 (1970).
313 Tonkelaar, W. A. M. den and G. Bergshoeff: Water Research 3, 31 (1969).
314 Truesdale, G. A., A. L. Downing and G. F. Lowden: J. Appl. Chem. 5, 53 (1955).
315 Turkolmez, S.: Vom Wasser 31, 153 (1964); Wasser/Luft/Betr. 9, 95 (1965).
316 Umbreit, G. R., R. E. Nygren and A. J. Testa: Pittsbgh. Conference Abstr. Paper Nr. 282 (1970).
317 Unger, U.: Gas/Wasserfach 112, 256 (1971).
317a Veith, G. D. and G. F. Lee: Water Research 4, 265 (1970).
318 Viehl, K.: Ges. Ing. 74, 123 (1953).
319 Vogel, H.: Sulfite Pulp Waste Effluents, Basel 1948.
320 van der Emde, W. and R. Kayser: Gas/Wasserfach 106, 1337 (1965).
321 Wagner, R.: Europ. Technicon Sympos. Paris 1966.
321a Wagner, R.: Chem. Abstr. 70, 120 907 (1966).
322 Wantschura, R.: Gas/Wasserfach 110, 711 (1969).
323 Waters, J. L., J. N. Little and D. F. Horgan: J. Chromat. Science 7, 293 (1969).
324 Webster, H. L. and J. Halliday: Analyst 84, 552 (1959).
325 Weil, L. and K. E. Quentin: Gas/Wasserfach 111, 26 (1970).
326 Weil, L. and K. E. Quentin: Z. Wasser- u. Abwasserforschg. 3, 67 (1970).

327   Weil, L. and K. E. Quentin: Gas/Wasserf. 112, 184 (1971).
328   Weil, L. and K. E. Quentin: Vortragsveröff. 1971, Haus d. Tech., Essen.
329   Weiss, C. M. and J. H. Gakstetter: J. Water Poll. Contr. Fed. 36, 240 (1964).
330   Wells, W. N.: Water/Sewage Works 117, 123 (1970).
331   Wernet, J. and K. Wahl: Z. anal. Ch. 251, 373 (1970).
332   Wexler, A. S.: Anal. Chemistry 35, 1936 (1963).
333   Wickbold, R.: Fette/Seifen/Anstrichm. 57, 164 (1955).
334   Wickbold, R.: Vom Wasser 33, 229 (1966).
335   Wickbold, R.: Fette/Seifen/Anstrichm. 70, 688 (1968).
335a  Wickbold, R.: Vortragsveröff. Haus d. Techn., Essen 1971.
336   Williams, J. L. and H. D. Graham: Anal. Chemistry 36, 1345 (1964).
337   Wildbrett, G., A. Ganz and F. Kiermeier: Z. Lebensm.-Unters. Forschg. 124, 69 (1964).
338   Wilder, E. T.: J. Amer. Water Works Assoc. 60, 827 (1968).
339   Winkler, L.: Z. anal. Ch. 41, 290 (1902).
340   Winter, G. D. and A. Ferrari: Residue Review 5 (1963).
341   Wolter, D.: Fortschr. Wasserchem. 7, 195 (1967).
342   Young, H. Y.: Bull. Envir. Contam. Toxicol. 2, 243 (1967); Z. anal. Ch. 246, 160 (1969).
343   Young, I. C., W. Garner and J. W. Clark: Anal. Chemistry 37, 784 (1965).
344   Zahner, R.: Vom Wasser 29, 172 (1962).
345   Zaleiko, N. S. and A. H. Molof: Proceed. 9th Nat. Anal. Inst. Sympos. Houston 1963.
346   Zigler, M. G. and W. F. Phillips: Envir. Sci. Technol. 1, 65 (1967).

# INDEX